IN SILICO GENOMICS AND PROTEOMICS: FUNCTIONAL ANNOTATION OF GENOMES AND PROTEINS

IN SILICO GENOMICS AND PROTEOMICS: FUNCTIONAL ANNOTATION OF GENOMES AND PROTEINS

NICOLA MULDER

AND

ROLF APWEILER

EDITORS

Nova Science Publishers, Inc.

New York

For permission to use material from this book please contact us:
Telephone 631-231-7269; Fax 631-231-8175
Web Site: http://www.novapublishers.com

NOTICE TO THE READER

The Publisher has taken reasonable care in the preparation of this book, but makes no expressed or implied warranty of any kind and assumes no responsibility for any errors or omissions. No liability is assumed for incidental or consequential damages in connection with or arising out of information contained in this book. The Publisher shall not be liable for any special, consequential, or exemplary damages resulting, in whole or in part, from the readers' use of, or reliance upon, this material.

This publication is designed to provide accurate and authoritative information with regard to the subject matter covered herein. It is sold with the clear understanding that the Publisher is not engaged in rendering legal or any other professional services. If legal or any other expert assistance is required, the services of a competent person should be sought. FROM A DECLARATION OF PARTICIPANTS JOINTLY ADOPTED BY A COMMITTEE OF THE AMERICAN BAR ASSOCIATION AND A COMMITTEE OF PUBLISHERS.

Library of Congress Cataloging-in-Publication Data
In silico genomics and proteomics : functional annotation of genomes and proteins / Nicola Mulder (editor).
    p. cm.
Includes index.
ISBN 1-59454-995-8
1. Genomics. 2. Proteomics. 3. Nucleotide sequence. I. Mulder, Nicola.
QH450.I5   2006
572.8'6--dc22   2006000417

Published by Nova Science Publishers, Inc. ✢ New York

Contents

Part A: Protein Annotation Tools

In: In Silico Genomics and Proteomics
Editors: N. Mulder and R. Apweiler, pp. 3-9

ISBN 1-59454-995-8
© 2006 Nova Science Publishers, Inc.

Chapter 1

Genomics and the Genome Era

Nicola J. Mulder and Rolf Apweiler

European Bioinformatics Institute, Wellcome Trust Genome Campus, Hinxton, Cambridge, UK, CB10 1SD

Abstract

The advancement of DNA sequencing technologies has resulted in a movement away from the single gene, and towards a whole genome focus. Previously DNA sequencing was expensive, slow and tedious, but new technologies have been developed that facilitate the relatively cheap and rapid generation of raw sequence from whole bacterial genomes. The sequencing of the human genome was also accelerated in the final phases as a result of new sequencing methods. However, generating the raw sequence is just the start, and tools are required to convert it into useful biological knowledge. This chapter highlights the new advances in DNA sequencing, and summarises the process of genome and protein annotation. The remaining chapters in the book describe the tools required to achieve this, and provide examples within the context of specific genomes.

Introduction

The first DNA sequencing method was developed in 1971 (Wu and Taylor 1971). This, and subsequent advances in sequencing technologies were a major break through in molecular biology, and opened the doors to whole new areas of exploration. New DNA sequencing protocols and instrumentation have been developed, facilitating an increase in the generation of raw sequence data. This has resulted in the emergence of genome sequencing centres that are releasing whole bacterial genome sequences almost weekly. However the flagstone event in the history of sequencing is undoubtedly the sequencing of the human genome. A race between the private and public domains culminated in the release of the first draft of the human genome by both parties in early 2001. This was an important milestone in genomics and promised significant consequences for science and medicine.

Once raw genome sequence data is generated, it needs to be computationally processed to convert it into a more useful form. Biologists are not interested in the sequence itself, but the information derived from the sequence, such as the features it encodes (Rouze *et al.*, 1999). This chapter highlights the new advances in DNA sequencing, and describes the process of genome and protein annotation, which converts raw sequence into biological knowledge. It summarises the following chapters and the potential applications and value of genomic information.

Genome Sequencing

DNA sequencing can be broken down into a number of individual steps, including sample collection, isolation and preparation, sequencing, assembly and finishing. Each of these steps has costs associated with it, and particular protocols or instrumentation required. Therefore, to improve sequencing technologies, it is necessary to improve one or ideally all of these steps. Cost is particularly important for each step, and can determine what genomes can be sequenced (Chan, 2005).

The first methods of DNA sequencing were laborious and involved terminal end-labelling, DNAse digestion, and 2-D electrophoresis on cellulose acetate and DEAE cellulose paper (Wu and Taylor 1971). The Sanger chain termination sequencing technique, published in 1977 (Sanger *et al.*, 1977), increased the efficiency and decreased the costs associated with sequencing. This was followed by the development of automated sequencing methods (Smith *et al.*, 1986; Hunkapiller *et al.*, 1991), which reduced the cost of sequencing to US\$ 1 per bp, and provided the capacity to sequence nearly 100,000 bp per day (Chan, 2005). This led the way for the sequencing of the Haemophilus influenza 1.8 megabase pair genome (Fleischmann *et al.*, 1995). In 1998, the ABI Prism® 3700 DNA Analyzer was introduced, which facilitated the sequencing of 900,000 bp per day. As a result, the cost of sequencing was reduced to US\$ 0.1 per bp, and it has since dropped 100-fold (Chan, 2005). Another important factor is read length, which has a significant impact on sequencing costs and the quality of the resulting data. Read length can be described as the contiguous length of sequences that can retrieved from a sequencing reaction (Chan, 2005). The first sequencing method of Wu and Taylor (1971), gave only 12 bp of sequence, while the Sanger method (Sanger *et al.*, 1977) gave reads of up to 300 bp. New capillary methods provide read lengths of about 1000 bp (Chan, 2005).

The development of new sequencing technologies, increase in read lengths, and reduction in price has led the way for larger genome sequencing projects, including the human and, more recently, other mammalian genomes. New genome sequencing centres have emerged that have the space, equipment and personnel to enable large-scale sequencing. These centres employ one or more different sequencing strategies, such as hierarchical shotgun or whole-genome shotgun sequencing, and fragments are usually cloned into bacterial artificial chromosomes for sequencing. However, despite advances in technologies, the cloning, sample preparation and sequencing is still expensive, and the estimated cost for producing a draft genome sequence the size of the human genome is US\$ 24 million. There are also limitations of current technologies employed by sequencing centres, for example in read length and the time taken to generate the reads. We are therefore still some way off from routine sequencing of the genomes of whole human populations, and even more advanced sequencing methods are required to achieve this goal with a reasonable price tag (Chan, 2005).

Genome Annotation

DNA sequencing machines churn out raw sequence in the form of chromatograms, and the sequence needs to be converted into a more biologically useful form using Bioinformatics tools. Software has been developed to read sequences from chromatograms, remove vector and other contaminating sequence, and facilitate base calling. Once the sequences for each chromatogram have been retrieved, they need to be assembled to generate the full sequence, and again, Bioinformatics tools are available for this.

The processing that follows sequence assembly includes gene prediction and annotation of the predicted proteins, collectively known as genome annotation. Genome annotation can loosely be divided into structural annotation, whereby features on a sequence are identified, and functional annotation, in which the structural features are functionally characterised (Rouze et al., 1999).

Structural Annotation

The first major step in the process of genome annotation is to identify features on the DNA sequence, which divides it into structural units. These features include coding and non-coding genes, promoters, regulatory regions, untranslated regions, repeats, etc. Genes and their predicted proteins are usually what the biologist is most interested in, and many of the features described above are related to the genes in providing their structure, expression and function (Rouze et al., 1999). Gene prediction programmes use the feature information and the presence of promoter regions, transcription start and stop sites and other characteristics of genes to identify them on a sequence. Many prediction programs also use sequence similarity to other genes or to expressed sequence tags (ESTs) or cDNAs to identify genes. Gene prediction for prokaryotic sequences is relatively straight forward, due to their short and simple gene structure, however eukaryotic genes are more complex and difficult to predict, particularly due to the presence of mRNA splicing. Eukaryotic genes often span vast regions of a DNA sequence and can be characterised by the presence of introns and exons. The major issues introduced by splicing are the fact that the splice sites are not always conserved and nucleotides at positions away from splice sites can affect splicing, and that exons can be small and sometimes non-coding, and thus difficult to identify (Mount, 2000). The availability of EST or cDNA sequences provides assistance in predicting introns and exons when the transcripts are mapped onto the genome sequence. Therefore, as the number of eukaryotic cDNA sequences in the database increases, so will the capacity for accurate gene prediction increase.

Functional Annotation

The functional annotation of predicted features on a genome sequence is the application of biological relevance to the feature. Among the different feature types, genes are the most studied, and the annotation of gene products is the main focus of this book. The most accurate method for the annotation of proteins is manual curation by experienced biologists. UniProtKB/Swiss-Prot (Bairoch *et al.*, 2005) entries are manually curated to a high standard and these protein entries are used as the gold standard for annotating uncharacterised proteins. However, manual curation is slow, and uncharacterised predicted proteins are entering the database at an exponential rate.

There are a number of different Bioinformatics tools available for protein annotation, but the most important breakthrough was the development of the basic local alignment search tool, BLAST (Altschul *et al.*, 1990). Sequence similarity searches usually constitute the first step in annotation, and, as mentioned above, are also used in gene prediction. If a query sequence has a strong similarity to a previously annotated sequence in the database, then the annotation of the query sequenced can be predicted. However, this relies on a related sequence being in the database, and preferably being annotated. Additional methods for protein annotation were developed using protein signatures. These are motifs that characterise a protein family or domain and can be predicted in new sequences. The InterPro database (Mulder *et al.*, 2005) integrates a number of public protein signature databases into a single resource, and is used by most of the genome sequencing centres for automatic annotation of gene products. Sequence clustering can also be used in a similar way for annotation, and both these methods, as well as manual curation methods, are described in chapters in this book.

Functional annotation of proteins can occur at different levels. The kind of functional information biologists require includes taxonomy of the organism the protein is from, tissue, molecular function, pathway, etc. The biological context of the protein in the cell is becoming increasingly important as whole genomes become available, and the pathways or networks the protein is involved in can also help with function prediction. Protein-protein interaction and biological pathway databases have emerged with the generation of protein interaction data from large-scale experiments. These new databases are being used for protein annotation and are described later in this book.

Genome Annotation Projects

The genome sequencing centres have developed pipelines for the annotation of sequences generated by their sequencing machines. In addition, specialised databases focussing on model organisms have developed for the annotation of these genomes, and these databases have also created their own annotation pipelines. These pipelines commonly use the tools described above, and results are usually presented in a user-friendly graphical interface. Some examples of genome annotation pipelines and genome browsers are Artemis (Berriman and Rutherford 2003), Ensembl (Hubbard *et al.*, 2005), VEGA (Ashurst *et al.*, 2005) and PEDANT (Frishman *et al.*, 2001). An example of a specialised genome database is

PlasmoDB (Bahl *et al.*, 2002), which provides a resource for *Plasmodium* genomics. These, as well as the microbial genome resource at NCBI and the Integr8 genome reviews and proteome analysis database are described in detail in individual chapters in this book.

Conclusion

Genome annotation is ongoing and will improve as new annotation tools are developed. In addition, as new genomes are sequenced, new protein families will emerge, and related genes can be annotated. If one or more members of a protein family are experimentally characterised, then the annotation can be inferred on other members. This is how manual UniProtKB/Swiss-Prot annotation is achieved, using sequence similarity searches and InterPro. The genome sequence and its annotation are valuable resources to the scientific community, and availability of annotation for sequence data provides the ability to mine the data more efficiently and thus biologists using the data can plan their experiments better. The experiments, in turn, provide biological validation for the predicted annotation (Rouze *et al.*, 1999). In addition, new laboratory techniques are being developed to facilitate large-scale experiments, such as microarrays or proteomics experiments, which will generate vast amounts of new experimental data to back up annotations. These validated protein functions are vital for scientific innovation, and are required for combating infectious diseases and for drug discovery. An understanding of the process of pathogenesis or how genes and processes are related to diseases is essential for modern medical and scientific research.

Genome sequences and annotations are not only valuable within the context of a single organism, but also for *in silico* comparative genomics or proteomics. Sequence similarity searches are small-scale comparative experiments and assist with protein function prediction, but large-scale comparative genomics can also aid function prediction through gene order and genomic context information, as well as providing data on the evolution of different organisms and the movement of genomic sequences through gene or genome duplication, horizontal transfer or recombination. The presence or absence of a genomic region in an organism may not be recognised without the comparison with other related organisms. New Bioinformatics tools have developed for comparative genomics or proteomics, some of which are described here.

This book aims to bring together all the disparate resources and tools that have been developed for processing and analysis of genomic data. The first part of the book describes the tools that have been developed for annotation, with an emphasis on the functional annotation of the gene products, since this is the point from which most Bioinformaticists and bench biologists work. The second part of the book describes specific genome annotation projects and comparative genomics and proteomics tools. There are a number of different annotation pipelines used by genome sequencing centres, for example, but it is unclear what the advantages and disadvantages of each are. Here we provide a resource for identifying many of the known tools and pipelines and similarities and differences between them.

References

Altschul SF, Gish W, Miller W, Myers EW and Lipman DJ. (1990) Basic local alignment search tool. *Journal of Molecular Biology,* 215(3): 403-410.

Ashurst JL, Chen CK, Gilbert JG, Jekosch K, Keenan S, Meidl P, Searle SM, Stalker J, Storey R, Trevanion S, Wilming L, Hubbard T. (2005) The Vertebrate Genome Annotation (Vega) database. *Nucleic Acids Research,* 33: D459-465.

Bahl A, Brunk BP, Coppel RL, Crabtree J, Diskin SJ, Fraunholz MJ, Grant G, Gupta D, Huestis RL, Kissinger JC, Labo P, Li L, McWeeney SK, Milgram AJ, Roos DS, Schug J and Stoeckert CJ Jr. (2002) PlasmoDB: The *Plasmodium* genome resource. An integrated database providing tools for accessing and analyzing mapping, expression and sequence data (both finished and unfinished). *Nucleic Acids Research,* 30: 87-90.

Bairoch A, Apweiler R, Wu CH, Barker WC, Boeckmann B, Ferro S, Gasteiger E, Huang H, Lopez R, Magrane M, Martin MJ, Natale DA, O'Donovan C, Redaschi N and Yeh LS (2005) The Universal Protein Resource (UniProt). *Nucleic Acids Research,* 33(1), D154-159.

Berriman M. and Rutherford K (2003) Viewing and annotating sequence data with Artemis. *Briefings in Bioinformatics,* 4(2): 124-132.

Chan EY. (2005) Advances in sequencing technology. *Mutation Research,* 573: 13-40.

Fleischmann RD, Adams MD, White O, Clayton RA, Kirkness EF, Kerlavage AR, Bult CJ, Tomb JF, Dougherty BA, Merrick JM, *et al.* (1995) Whole-genome random sequencing and assembly of *Haemophilus influenzae* Rd. *Science,* 269: 496–512.

Frishman D, Albermann K, Hani J, Heumann K, Metanomski A, Zollner A and Mewes HW (2001) Functional and structural genomics using PEDANT. *Bioinformatics,* 17: 44-57.

Hubbard T, Andrews D, Caccamo M, Cameron G, Chen Y, Clamp M, Clarke L, Coates G, Cox T, Cunningham F, Curwen V, Cutts T, Down T, Durbin R, Fernandez-Suarez XM, Gilbert J, Hammond M, Herrero J, Hotz H, Howe K, Iyer V, Jekosch K, Kahari A, Kasprzyk A, Keefe D, Keenan S, Kokocinsci F, London D, Longden I, McVicker G, Melsopp C, Meidl P, Potter S, Proctor G, Rae M, Rios D, Schuster M, Searle S, Severin J, Slater G, Smedley D, Smith J, Spooner W, Stabenau A, Stalker J, Storey R, Trevanion S, Ureta-Vidal A, Vogel J, White S, Woodwark C and Birney E (2005) Ensembl 2005. *Nucleic Acids Research,* 33: D447-453.

Hunkapiller T, Kaiser RJ, Koop BF and Hood, L. (1991) Large-scale and automated DNA sequence determination. *Science,* 254: 59–67.

Mount S. (2000) Genomic sequence, splicing and gene annotation. *American Journal of Human Genetics,* 67: 788-792.

Mulder, N. J., Apweiler, R., Attwood, T. K., Bairoch, A., Bateman, A., Binns, D., Bradley, P., Bork, P., Bucher, P., Cerutti, L., Copley, R., Courcelle, E., Das, U., Durbin, R., Fleischmann, W., Gough, J., Haft, D., Harte, N., Hulo, N., Kahn, D., Kanapin, A., Krestyaninova, M., Lonsdale, D., Lopez, R., Letunic, I., Madera, M., Maslen, J., McDowall, J., Mitchell, A., Nikolskaya, A. N., Orchard, S., Pagni, M., Ponting, C. P., Quevillon, E., Selengut, J., Sigrist, C. J., Silventoinen, V., Studholme, D. J., Vaughan, R. and Wu, C. H. (2005) InterPro, progress and status in 2005. *Nucleic Acids Research,* 33(1), D201-5.

Rouze P, Pavy N and Rombauts S. (1999) Genome annotation: which tools do we have for it? *Current Opinion in Plant Biology,* 2: 90-95.

Sanger F, Nicklen S and Coulson AR. (1977) DNA sequencing with chain-terminating inhibitors. *Proceedings of the National Academy of Sciences U.S.A.* 74: 5463–5467.

Smith LM, Sanders JZ, Kaiser RJ, Hughes P, Dodd C, Connell CR, Heiner C, Kent SB, Hood LE. (1986) Fluorescence detection in automated DNA sequence analysis. *Nature,* 321: 674–679.

Wu R, and Taylor E. (1971) Nucleotide sequence analysis of DNA II. Complete nucleotide sequence of the cohesive ends of bacteriophage lambda DNA. *Journal of Molecular Biology,* 57: 491–511.

In: In Silico Genomics and Proteomics
Editors: N. Mulder and R. Apweiler, pp. 11-23

ISBN 1-59454-995-8
© 2006 Nova Science Publishers, Inc.

Chapter 2

Sequence Clustering as a Method of Protein Functional Annotation

R. Petryszak and P. Kersey

European Bioinformatics Institute, Wellcome Trust Genome Campus, Hinxton,
Cambridge, UK, CB10 1SD

Abstract

This chapter begins with an overview of clustering as a useful technique for analysing
data-rich problems, which is followed by a discussion of selected automatic methods for
protein sequence clustering. The chapter ends with a brief overview of applications of
sequence clustering to phylogenetic analysis and protein classification, and a note on the
importance of updating and maintainability of protein sequence clustering databases.

1. Introduction to Sequence Clustering Algorithms

Clustering is a method for analysing data-rich problems by grouping data items into
similarity classes. One motivation behind clustering may be to reduce the problem space, so
that one need only consider a common characteristic or a representative member of each
class, rather than the full data set. Clustering is also a convenient tool for characterising
individual data items through their cluster membership. Finally, clustering can be used for
data mining, e.g. attempting to characterise an unknown data set by tracking its clustering
patterns. The similarity measure and thus the set of characteristics shared by all members of a
given class are defined by the problem domain, and different similarity measures may yield
different and possibly equally valid classifications of that domain.

The basic clustering steps are as follows:

- Parameterise the problem domain data set so that a notion of similarity can be applied to discriminate between data items, e.g. model proteins as amino acid sequences.
- Define a distance metric, which can be used as a measure of similarity between two (sets of) data items, e.g. the number of adjacent amino acids shared by two protein sequences.
- Use the above distance metric to classify the problem domain data set into groups of similar data items, e.g. derive clusters of protein sequences whose members share sequence similarity with at least one other member of the same cluster.
- Validate (and apply) the classification of the problem domain data set. In this, arguably most complex, step the derived classification is used to gain new insights into the problem domain.

A number of clustering techniques exist, among which two major types can be distinguished:

Non-Hierarchical Clustering

A non-hierarchical clustering algorithm assigns each new data item to its closest (as defined by the similarity function) cluster. If no cluster is found above a certain similarity threshold, the algorithm creates a new cluster with that new data item as its sole member. The similarity threshold controls the average size and the amount of the clusters found. The optimal value of this threshold may need to be derived in conjunction with the domain knowledge by repeating the clustering step with different threshold values until a set of clusters is found that best characterises the problem domain data set. One inherent drawback of the basic algorithm is that the decisions on cluster assignment for new data items have to be made before all the available clusters are known. This problem can be addressed by splitting the basic process into two steps:

- first create singleton clusters for all data items which are not close enough to any cluster created so far (according to the chosen similarity threshold);
- then assign all the yet unassigned data items to their respective closest clusters created in the first step.

Graph Methods

Graph methods draw on ideas from graph theory when choosing clusters to join. As exemplified in Figure 1, the similarity matrix can be represented as a graph in which vertices correspond to data items and edges represent pairwise similarities between data items. The length of an edge connecting two nodes can represent the degree of similarity between the corresponding data. Alternatively, weights can be attached to edges to represent the level of similarity between data items they connect. An example of graph-based non-hierarchical clustering is embodied in the Markov Clustering (MCL) algorithm (Enright *et al.*, 2002) in which weights are transformed into probabilities associated with a transition from one data item (node) to another within the graph. This matrix is passed through iterative rounds of matrix multiplication and matrix inflation (squaring each element of the matrix separately) until there is little or no net change in the matrix. The final matrix is then interpreted as the resultant set of clusters. The MCL algorithm is described in greater detail in the latter part of this chapter.

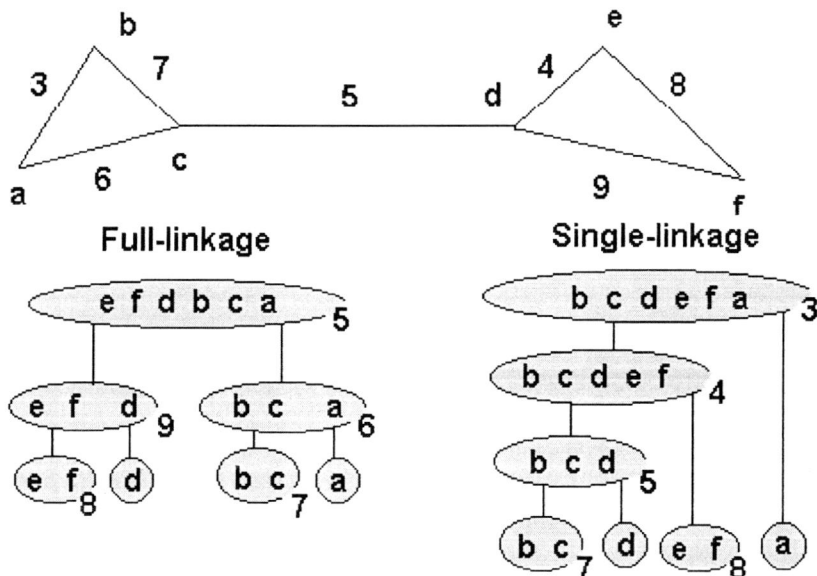

Note that the full-linkage algorithm yields more compact clusters than its single-linkage counterpart. Note also that the top cluster under the full-linkage would not be created if the requirement for members of a cluster to be similar to all other members of that cluster was adhered to. The similarity scores, assigned to the non-singleton clusters, have been designed for the purpose of illustration to uniquely identify the similarities that led to the creation of a given cluster.

Figure 1. The similarity matrix among data items a to f is visualized as a weighted graph, in which the data items are represented by the vertices and the similarities between them as edges. The weights assigned to edges correspond to similarity scores.

K-Means

The principle of k-clustering is that for an input set S and a fixed integer k, a partition of S into k subsets must be returned. K-means clustering (MacQueen, 1967) is the most common example of k-clustering. The goal of k-means clustering is to minimise dissimilarity among the members of each cluster while maximising this value between elements in different clusters. One example way of computing dissimilarity between a sequence x and a cluster C is by choosing a representative member of the cluster (a centroid[1]) and then, instead

---

[1] The *centroid* is a cluster member whose parameter values are the average of the parameter values of all members of that cluster.

of computing distances between x and all members of C, compute only the distance between x and the centroid of C.

The following steps are involved in k-means algorithm, given the set of data items, S, and the expected number of resulting clusters, k:

- Select k initial cluster centroids: c1, c2, ..., ck; This can be achieved for example by random partitioning of S into K clusters and then computing their centroids; or simply by random sampling of k elements.
- Repeat the following until the convergence is achieved:
- Assign each data item x in S to the cluster whose centroid is the nearest to x;
- For each cluster, re-compute its centroid based on its new member set.

For an example application of K-means algorithm, used for clustering genes based on their expression patterns, see Dhillon *et al.*, 2003.

Self-Organising Maps (SOMs, also Known as Kohonen Maps (Kohonen, 1990))

A SOM is a learning algorithm, implemented as a neural network[2], which classifies input data in an unsupervised manner. This approach manages to organise highly complex high-dimensional data (e.g. human speech) into a low-dimensional representation (e.g. 2-D graph) by putting similar entities geometrically close to each other. For an example of an application of SOM-based clustering to the analysis of gene expression data see Tamayo *et al.*, 1999.

Hierarchical Clustering

The hierarchical clustering yields a hierarchy of nested clusters, where larger clusters are created using a lower similarity threshold than their respective children clusters. Consequently, cluster sizes increase as the degree of similarity among their members decreases. There exist two main categories of hierarchical clustering: *agglomerative* and *inverse divisive* methods. The basic agglomerative method starts with N singleton[3] clusters (a.k.a. leaves) and joins them incrementally into increasingly large clusters then until the final hierarchy results.

The inverse divisive method achieves the opposite: start with one flat clustering and incrementally split clusters until N leaves emerge.

---

[2] A Neural Network is a biologically-inspired information-processing architecture, which consists of a network of nodes referred to as neurons. These nodes, like biological neurons, receive input from other neurons and produce an output if a combined value of its inputs exceeds a certain threshold. A neural network, given enough example data, can learn what a specific pattern looks like (training mode), and then categorise previously unseen patterns by whether they are similar to the pattern it was trained to recognise or not (testing mode).

[3] A singleton cluster contains one data item as its sole member.

Agglomerative Methods

The process of transforming N singleton clusters into the final hierarchy relies on the presence of N x N pairwise similarity matrix between all data items. An example of an agglomerative method, called *matrix method*, updates the similarity matrix at each cluster joining so that the matrix always represents the similarity between all current clusters[4]. As soon as two clusters C_1 and C_2 are joined into a parent cluster C_p, the data for C_1 and C_2 in the similarity matrix are replaced by the set of similarity distances between C_p and all other parent-less clusters.

The key difference among the matrix methods lies in the way the similarity (distance) between the parent cluster C_p and an existing cluster C_x ($Sim(C_p, C_x)$) is calculated, based on the distances to Cx from C_1 and C_2 ($Sim(C_1, C_x)$ and $Sim(C_2, C_x)$ respectively).

In the *single-linkage* algorithm, $Sim(C_p, C_x)$ is the *lesser* of $Sim(C_1, C_x)$ and $Sim(C_2, C_x)$, whereas under the *full-linkage* principle the *greater* of $Sim(C_1, C_x)$ and $Sim(C_2, C_x)$ is chosen. As a consequence, in the case of a tightly coupled section of the similarity matrix (i.e. in which data items are similar to more than one other data item) the full-linkage algorithm will yield more compact clusters than the single-linkage method. An extreme case of the full-linkage algorithm can be envisaged that yields only clusters in which all members are similar to all other members of the respective cluster. In this case the full-linkage similarity distance calculation principle still applies, but similarities can no longer be considered in isolation. Namely, only if similarities between that new data item and all members of a given cluster exist, can that item be made a member of that cluster (see Figure 1). Finally, the *average-linkage* algorithm can also be used, in which a number of mathematical formulae are available for quantifying inter-cluster distance, e.g. an *arithmetic mean, a square root of an arithmetic mean of squares, a geometric mean* and a *harmonic mean* (Sasson *et al.*, 2002).

A less computationally-intensive version of single-linkage algorithm exists which does not require the re-calculation of distances between parent-less clusters at each step of the clustering process. Instead, the pairwise similarities (between data items) in the original matrix are first sorted by the similarity score, in descending order. Consequently, when this sorted list of similarities is traversed, stronger similarities are encountered before weaker ones. This way the cluster hierarchy can be built incrementally, based solely on the existing cluster membership of the proteins involved in a given similarity. In the most general case, when all similarity levels are considered, each new similarity S_{ab} (between proteins a and b) will cause a new parent cluster to be created, unless an existing cluster C_x (which, by definition, groups proteins at a level of similarity which is higher than S_{ab}) already contains data items a and b as members.

The following steps are taken when including the similarity S_{ab} in the existing cluster hierarchy:

- If a singleton cluster C_a does not exist, create one. Assign the highest level of similarity (between C_a and itself, i.e. identity) to C_a;
- Create a singleton C_b if necessary, as for item a;

- If cluster C_{xab} (member set of which includes a and b) exists, then discard S_{ab};
- If clusters C_{xa} (of which a is a member) and C_{yb} (of which b is member) exist, then create cluster C_{abxy} and make C_{xa} and C_{yb} its children.
- Otherwise, if only C_{xa} exists, create cluster C_{xab} and make C_{xa} and C_b its children.
- Otherwise, if only C_{xb} exists, create cluster C_{xab} and make C_{xb} and C_a its children.
- Otherwise create cluster C_{ab} and make C_a and C_b its children.

In all cases of new cluster creation, assign the similarity score of S_{ab} as the level of similarity of the newly created cluster.

This version of single-linkage algorithm is used by CluSTr database (Kriventseva *et al.*, 2001), which is described in more detail in the latter part of this chapter.

Divisive Algorithm

This algorithm is the converse of the agglomerative method, starting with one all-inclusive cluster and then splitting clusters until all data items are separate. At each step, all possible partitions of clusters are checked to find the one that would yield two clusters of minimum similarity (since we want the most dissimilar sets of items to be in separate clusters as soon as possible). However, since checking partitions of clusters on a large scale is not practicable, in practice heuristic rules are used to select the most promising partitions. One example heuristic for removing a partition C_p from cluster C_x is to first move to C_p the item which is the least similar to all other members of C_x, and then continue moving items from C_x to C_p as long as their average similarity to C_p is greater than to C_x. This algorithm will stop when no members of C_x are more similar to C_p than to C_x.

2. Applications of Clustering Algorithms in Protein Sequence Clustering

2.1. Protein Sequence Classification by Automatic and Manual Methods

Proteins can be classified into families, i.e. groups of proteins which are related by evolution. As a result of many large-scale genome sequencing and translation projects, the number of publicly available protein sequences is growing exponentially. Given the prohibitively high cost of manual curation, fully or semi-automatic methods of protein classification increasingly grow in prominence, through their high coverage at a reasonable level of precision. Examples of resources in which manual curation is used to refine families found by automatic classification procedure are InterPro (Mulder *et al.*, 2003), Pfam (Bateman *et al.*, 2004), Clusters of Orthologous Groups (COGs) (Tatusov *et al*, 2003) and PIR SuperFamily (Wu *et al.*, 2003). Fully automatic classification methods, on the other

---

[4] Note that the matrix of similarities between all data items automatically becomes the matrix of similarities between the singleton clusters created from those data items to seed to agglomerative clustering process.

hand, are used in e.g. CluSTr, UniRef (Apweiler *et al.*, 2004), ProtoMap (Yona *et al.*, 2000) and TribeMCL (Enright *et al.*, 2002).

2.2. Automatic Protein Classification by Sequence Clustering

One approach for automatic classification of proteins is clustering by sequence similarity. Such *sequence clustering* methods often derive similarity scores from sequence alignments (CluSTr, ProtoMap). The clustering procedure then uses those similarity scores to build clusters of proteins with a similarity over a particular threshold (e.g. single-linkage – see Figure 1).

The sequence clustering procedure can return either a set of clusters (CD-HIT (Li *et al.*, 2001), TribeMCL) or a cluster hierarchy (CluSTr, ProtoMap). In the latter case, each level of the hierarchy represents different degree of similarity between clustered sequences. Hierarchical clustering at all levels of similarity allows one to divorce decisions about what degrees of similarity would result in biologically meaingful clusters from the clustering process itself. After an initial hierachy of clusters is generated, other (divergent or even mutually contradictory) methods may be applied on that hierarchy to detect which clusters and similarity levels are biologically revealing.

Once families of homologous proteins are identified, annotation associated with well-characterised members of the family can be assigned to unknown family members. Due to a limited scope of this chapter only a short representative list of sequence clustering methods and applications is presented below.

CD-HIT

CD-HIT method of clustering stands for *Cluster Database at High Identity with Tolerance* and is used by e.g. UniRef. The program (cd-hit) takes a FASTA format sequence database as input and produces a set of 'non-redundant' (nr) representative sequences as output. In addition cd-hit outputs a cluster file, documenting the remaining members of the cluster represented by each nr sequence. Consequently, a reduction of the overall size of the database is achieved without removing any sequence information by only removing 'redundant' (i.e. highly similar) sequences. CD-HIT uses a 'longest sequence first' list removal algorithm to remove sequences above a certain identity threshold. Additionally, the algorithm implements a very fast heuristic to find high identity segments between sequences, and so can avoid many costly full alignments.

TRIBE-MCL

The TRIBE-MCL method provides rapid and accurate non-hierarchical clustering of protein sequences into families on a large scale. It relies on the Markov cluster (MCL) algorithm, previously developed for graph clustering using flow simulation, for the assignment of proteins into families based on a BLAST sequence similarity matrix. Enright *et al.* show that this approach does not suffer from the problems that normally hinder other protein sequence clustering algorithms, such as the presence of multi-domain proteins, the so-called 'promiscuous domains' (e.g. SH2 or WD40 - domains which share similar structure but through evolution have been adapted to divergent functions and therefore are present in many functional families), and fragmented proteins.

The TRIBE-MCL approach to protein clustering is fast and accurate, avoiding most of the problems described above. The algorithm first represents the sequence similarity matrix from the BLAST run as a weighted graph (as described earlier in the chapter). The weights (similarity scores) associated with edges are converted into probabilities associated with a transition from one protein to another within the graph. This matrix is passed through iterative rounds of matrix expansion[5] (the normal matrix product) and matrix inflation[6] (the Hadamard power of a matrix, i.e. taking powers entry-wise), followed by a scaling step. The scaling step is designed to render the resulting matrix stochastic again (i.e. the matrix elements correspond to probability values).

Eventually, iterating expansion and inflation results in the separation of the graph into different segments. There are no longer any paths between these segments and the collection of resulting segments is simply interpreted as a protein family clustering. The inflation value parameter of the MCL algorithm is used to control the granularity of these clusters. Inter-cluster paths represent either sequence similarity relationships due to multi-domain proteins or mere false positive similarity detections.

Many of the problems that normally hinder protein sequence clustering are eliminated by the MCL approach. Proteins possessing a promiscuous domain, which is present in many functionally unrelated proteins, are normally very difficult to cluster correctly. Promiscuous domains will connect a member of a given protein family to all members of that family and possibly to other protein families. Because these inter-family connections are still far fewer that intra-family connections, the algorithm gradually eliminates these inter-family similarities and detects functional protein families accurately. The algorithm requires no a priori knowledge of protein domains, and clusters proteins into families purely based on observed relationships through the entire similarity graph. Comparisons with SCOP (Murzin et al., 1995) and InterPro (see Enright et al., 2002) have illustrated MCL algorithm's ability to cluster proteins with different domain structures into distinct families, despite the fact that structural similarities are not readily detectable at the sequence level. The quality and accuracy of MCL clustering was also illustrated by the fact that the InterPro domain of almost all proteins matched the most frequently occurring domain combination of the cluster they belonged to.

CluStr[7]

The CluStr database is built using a fully automatic hierarchical clustering method to yield hierarchies of protein families, based on a similarity matrix. In order to compute the

---

[5] The expansion phase corresponds to computing random walks of 'higher length', which means random walks with many steps. It associates new probabilities with all pairs of nodes in the matrix. Since higher length paths are more common within clusters than between clusters, the probabilities associated with node pairs lying in the same cluster will, in general, be relatively large as there are many ways of going from one node to the other.

[6] The Inflation boosts the probabilities of intra-cluster walks and demotes inter-cluster walks, and is not the result of any prior knowledge of cluster structure.

[7] CluStr database classifies protein sequences from both the UniProt Knowledgebase (Apweiler R. et al., 2004) and IPI (Kersey P. J. et al., 2004), and at the time of writing covers around 700,000 protein sequences and 195 complete proteomes (including 11 eukaryotes).

matrix, firstly all-against-all pair-wise comparisons between protein sequences are computed using the Smith-Waterman algorithm, resulting in a set of similarities and their corresponding Smith-Waterman scores.

Secondly, a Monte-Carlo simulation is performed to assess a statistical significance of the Smith-Waterman scores, yielding Z-scores (Comet *et al., 1999*). It is the Z-scores that are used as a measure of sequence similarity in the similarity matrix.

Given the set of similarities and their associated z-scores, a single-linkage clustering procedure (for an example, see Figure 1 and Figure 2) takes place, which yields a hierarchy of clusters. When the hierarchy is traversed from children to parents, cluster sizes get progressively larger, and the corresponding Z-scores, and hence the similarity levels, get smaller.

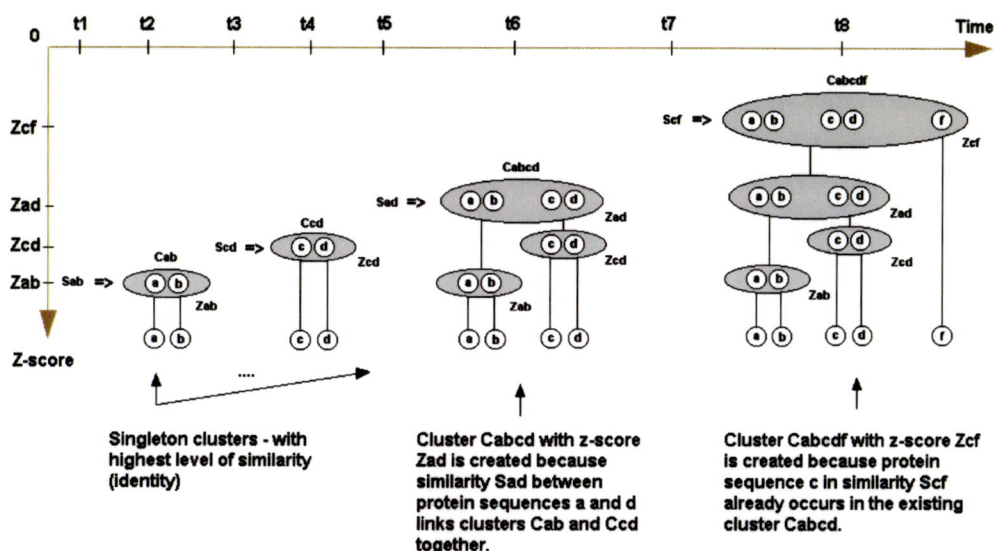

Figure 2. Single-linkage clustering procedure in CluSTr

This procedure first sorts protein sequence similarities by Z-score in descending order, and then creates singleton clusters for all proteins which take part in those similarities. The singletons are assigned the maximum level of similarity (identity) and are simply artifacts of the clustering procedure. Smaller clusters, with higher corresponding levels of similarity (Z-scores), are then merged into bigger clusters, with lower z-scores. Under the principle of single-linkage, a new similarity S_{ab} between sequences a and b at z-score Z_{ab} will result in merging of clusters C_a and C_b (created in previous steps, at a Z-score higher than Z_{ab}) into cluster C_{ab} if sequence a is a member of C_a and sequence b is a member of C_b.

C_{ab} is from then on referred to as a parent of C_a and C_b, and C_a and C_b as children of C_{ab}. The process of creating parent clusters continues until the set of sequence similarities is exhausted.

The resulting hierarchy of clusters is a binary forest (see Figure 3) in which each parent has only two children, but where a number of parent-less clusters (a.k.a. *ultimate predecessors*) may exist.

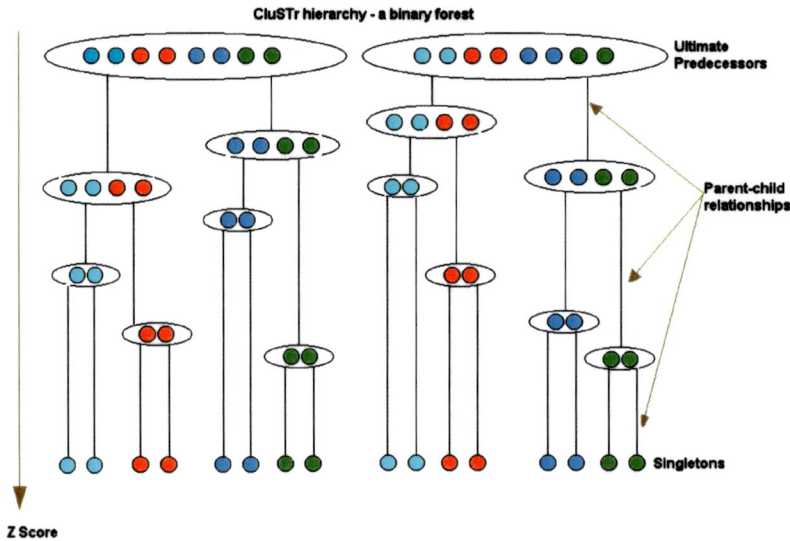

Figure 3. CluSTr hierarchy - a binary forest

The quality of CluSTr data has been verified in an automated annotation experiment (Petryszak *et al.*, 2005) involving predicting keywords, descriptions and comments, found in the KW, DE and CC lines of the UniProt Knowledgebase/SwissProt entries respectively, using standard data mining technology. In the course of the experiment, a comparison was made between the performance of predictive models (annotation rules) derived from the following data sources:

- InterPro
- InterPro and CluSTr data combined,
- InterPro and CluSTr data combined, but with Pfam signature hits removed.

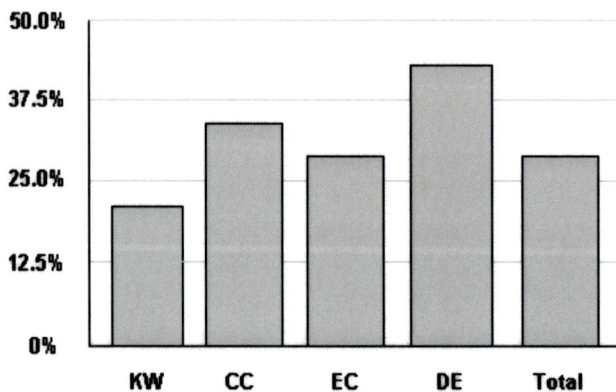

Figure 4. Performance improvement (the amount of annotation errors avoided) of the automated annotation predictions once the CluSTr data set is exploited in addition to traditional methods. Diagram distinguishes performance for keywords (KW line of the UniProt entry), comments (CC), EC numbers and descriptions (DE)

To make the results comparable, all tests were performed on the set of UniProt Knowledgebase/SwissProt proteins that are covered by CluSTr. The experiment involved a direct comparison of the precision and the coverage of the annotation rules derived from all three data sources. The results of the experiment show that, overall, the automated annotation predictions benefit largely by the inclusion of CluSTr as a data type. The recall rate increases slightly by 1.3%, while the impact on precision is noteworthy. More than 30% of all annotation errors can be avoided, as Figure 4 illustrates.

3. Applications of Sequence Clustering to Phylogenetic Analysis and Protein Classification

A common use of clustering techniques is the determination of genes that share common evolutionary ancestry. Two genes are considered to be orthologous if they arose from speciation; and paralogous if they arose from duplication. Identification of orthology and paralogy allows us to trace the lineage of individual genes (which, owing to lateral gene transfer, may differ from that of species), and aids the characterisation and analysis of complete genomes. It can also provide insights into the process of evolution. Accurate determination of orthology and paralogy is dependent on phylogenetic tree reconstruction, usually by methods such as maximum likelihood. In this method, assumptions about the behaviour of biological systems are used to increase accuracy, compared with the use of simple methods based on sequence distance. Such methods are, however, frequently very computationally expensive, and need to be applied to a set of genes already identified as related. Distance-based sequence clustering is thus still often used to identify families of related genes to which more sophisticated analysis can then be applied; for example, the HOGENOM database of homologous genes (Dufayard *et al.*, 2005) has been constructed in this way.

The COG database (Tatusov *et al.*, 2003) attempts to identify putative orthologues without the need to construct phylogenetic trees. COG uses a simple, non-hierarchical clustering technique based on the identification of reciprocal best matches between pairs of complete genomes, and the assembly of these pairwise matches into larger groups. The PIRSF database of homeomorphic families (Wu *et al.*, 2004) is also built according to a similar methodology: firstly, sequence similarity analysis is used to divide proteins into clusters; then, each cluster is subject to the application of additional (taxonomic and domain-based constraints); and then each resulting protein set is used to build Hidden Markov models that can be used to classify new protein sequences into the appropriate family.

The identification of potential orthologues can be usefully visualised through the use of genome browsing software, highlighting the extent of synteny between organisms (for example, see http://www.ensembl.org (Birney *et al.*, 2004), where potential orthologues in a number of higher eukaryotic species identified using the TRIBE-MCL algorithm can be visualised; and http://www.ebi.ac.uk/integr8 (Kersey *et al.,* 2004), where the CluSTr database provides this information for all species with completely deciphered genomes).

4. Update and Maintainability of Protein Sequence Cluster Data

In order to maximise the value of a clustering database, its data needs to be kept up-to-date as much as possible. Given the high growth of available protein sequences, regular updates to clustering resources can consume a prohibitive amount of computational power. A strategy for dealing with this problem has been factored into the design of the CluSTr database (see http://www.ebi.ac.uk/clustr). In CluSTr, it is the pairwise similarity calculations, rather than the clustering procedure itself, that incur the higher computational cost. However, each pairwise similarity is quantified using a statistical measure (Z-score) that is dependent only on the compared sequences and not on the size and composition of the sequence database. Consequently, the Z-scores for all existing sequences are kept and only "new-against-new" and "new-against-existing" similarity calculations are required in each CluSTr update.

This strategy enables the CluSTr database to be kept up-to-date for a continuously growing set of completely deciphered proteomes and to do so in synchrony with the fortnightly UniProt (see http://www.uniprot.org) update cycle.

References

Apweiler R., Bairoch A., Wu C.H., Barker W.C., Boeckmann B., Ferro S., Gasteiger E., Huang H., Lopez R., Magrane M., Martin M.J., Natale D.A., O'Donovan C., Redaschi N., Yeh L.L. (2004) UniProt: the Universal Protein knowledgebase. *Nucleic Acids Res.*, 32, D115-D119.

Bateman A., Coin L., Durbin R., Finn R. D. , Hollich V., Griffiths-Jones S., Khanna A., Marshall M., Moxon S., Sonnhammer E. L. L., Studholme D. J., Yeats C. and Eddy S. R. (2004) The Pfam Protein Families Database. *Nucleic Acids Research*, 32, D138-D141.

Birney E, Andrews D, Bevan P, Caccamo M, Cameron G, Chen Y, Clarke L, Coates G, Cox T, Cuff J, Curwen V, Cutts T, Down T, Durbin R, Eyras E, Fernandez-Suarez XM, Gane P, Gibbins B, Gilbert J, Hammond M, Hotz H, Iyer V, Kahari A, Jekosch K, Kasprzyk A, Keefe D, Keenan S, Lehvaslaiho H, McVicker G, Melsopp C, Meidl P, Mongin E, Pettett R, Potter S, Proctor G, Rae M, Searle S, Slater G, Smedley D, Smith J, Spooner W, Stabenau A, Stalker J, Storey R, Ureta-Vidal A, Woodwark C, Clamp M, Hubbard T. (2004) Ensembl 2004. *Nucleic Acids Res.*, 32(Database issue): D468-D470.

Dhillon I. S., Marcotte E. M., Roshan U. (2003) Diametrical clustering for identifying anti-correlated gene clusters. *Bioinformatics*, 19, 1612-1619.

Dufayard J.-F., Duret L., Penel S., Gouy M., Rechenmann F, and Perrière G. (2005) Tree pattern matching in phylogenetic trees: automatic search for orthologs or paralogs in homologous gene sequence databases. *Bioinformatics*, doi:10.1093/bioinformatics/bti325

Enright A.J., Van Dongen S., Ouzounis C.A. (2002) An efficient algorithm for large-scale detection of protein families. *Nucleic Acids Res.*, 30, 1575-1584.

Kersey P, Bower L, Morris L, Horne A, Petryszak R, Kanz C, Kanapin A, Das U, Michoud K, Phan I, Gattiker A, Kulikova T, Faruque N, Duggan K, Mclaren P, Reimholz B, Duret

L, Penel S, Reuter I, Apweiler R. (2005) Integr8 and Genome Reviews: integrated views of complete genomes and proteomes, *Nucleic Acids Res.*, 33(Database Issue): D297-D302.

Kohonen, T. (1990). The self-organizing map. *Proceedings of the IEEE*, 78(9), 1464-1480.

Li W., Jaroszewski L., Godzik (2001) A. Clustering of highly homologous sequences to reduce the size of large protein databases. *Bioinformatics*, 17, 282-283.

Kriventseva E.V., Fleischmann W., Zdobnov E.M., Apweiler R. (2001) CluSTr: a database of clusters of SWISS-PROT+TrEMBL proteins. *Nucleic Acids Res.*, 29(1), 3-6.

MacQueen J. B. (1967) Some Methods for classification and Analysis of Multivariate Observations, Proceedings of 5-th Berkeley Symposium on Mathematical Statistics and Probability, Berkeley. *University of California Press*, 1, 281-297

Mulder N.J., Apweiler R., Attwood T.K., Bairoch A., Barrell D., Bateman A., Binns D., Biswas M., Bradley P., Bork P., Bucher P., Copley R.R., Courcelle E., Das U., Durbin R., Falquet L., Fleischmann W., Griffiths-Jones S., Haft D., Harte N., Hulo N., Kahn D., Kanapin A., Krestyaninova M., Lopez R., Letunic I., Lonsdale D., Silventoinen V., Orchard S.E., Pagni M., Peyruc D., Ponting C.P., Selengut J.D., Servant F., Sigrist C.J.A., Vaughan R., Zdobnov E.M. (2003) The InterPro Database, 2003 brings increased coverage and new features. *Nucleic Acids Res.*, 31, 315-318.

Murzin A. G., Brenner S. E., Hubbard T., Chothia C. (1995). SCOP: a structural classification of proteins database for the investigation of sequences and structures. *J. Mol. Biol.*, 247, 536-540.

Petryszak R, Kretschmann E., Wieser D., Apweiler R. (2005). The predictive power of the CluSTr database. *Bioinformatics*, 21, 3604 - 3609.

Sasson O., Linial N., Linial M. (2002) The metric space of proteins – comparative study of clustering algorithms. *Bioinfiormatics*, 18, S14-S21.

Tamayo P., Slonim D., Mesirov J., Zhu Q., Kitareewan S., Dmitrovsky E., Lander E. S., Golub R. T. (1999) Interpreting patterns of gene expression with self-organizing maps: methods and application to hematopoietic differentiation. *Proc. Natl. Acad. Sci. USA*, 96, 2907-2912.

Tatusov R.L., Fedorova N.D., Jackson J.D., Jacobs A.R., Kiryutin B., Koonin E.V., Krylov D.M., Mazumder R., Mekhedov S.L., Nikolskaya A.N., Rao B.S., Smirnov S., Sverdlov A.V., Vasudevan S., Wolf Y.I., Yin J.J., Natale D.A. (2003) The COG database: an updated version includes eukaryotes. *BMC Bioinformatics*, 4, 41.

Wu C.H., Yeh L.L., Huang H., Arminski L., Castro-Alvear J., Chen Y., Hu Z., Ledley R.S., Kourtesis P., Suzek B.E., Vinayaka C.R., Zhang J., Barker W.C. (2003) The Protein Information Resource. *Nucleic Acids Res.*, 31, 345-347.

Wu CH, Nikolskaya A, Huang H, Yeh LS, Natale DA, Vinayaka CR, Hu ZZ, Mazumder R, Kumar S, Kourtesis P, Ledley RS, Suzek BE, Arminski L, Chen Y, Zhang J, Cardenas JL, Chung S, Castro-Alvear J, Dinkov G, Barker WC. (2004) PIRSF: family classification system at the Protein Information Resource. *Nucleic Acids Res.*, 32(Database issue): D112-D114.

Yona G., Linial N., Linial M. (2000) ProtoMap: automatic classification of protein sequences and hierarchy of protein families. *Nucleic Acids Res.*, 28, 49-55.

In: In Silico Genomics and Proteomics
Editors: N. Mulder and R. Apweiler, pp. 25-36

ISBN 1-59454-995-8
© 2006 Nova Science Publishers, Inc.

Chapter 3

Function Prediction with Protein Signatures and InterPro

Nicola J. Mulder and Rolf Apweiler

European Bioinformatics Institute, Wellcome Trust Genome Campus, Hinxton,
Cambridge, UK, CB10 1SD

Abstract

The multitude of new sequences generated from genome sequencing projects cannot be annotated at the same rate as they are produced. Therefore reliable methods for automatic functional classification of proteins are a necessity and many such methods are described in this book. There are a number of groups that use the approach of identifying motifs or domains found in previously characterised protein families. These conserved regions are predicted to be indicators of protein function. The two main approaches for identifying and describing these regions are sequence clustering and protein signatures. Protein signatures can be produced using different techniques such as regular expressions, profiles or hidden Markov models, and are more sensitive and effective than sequence similarity searches. Protein signatures are the tools of choice for protein annotation due to their accuracy, and their ability to include distantly related family members. There are a number of databases in the public domain that use these techniques, including Pfam, PROSITE, PRINTS, SMART, TIGRFAMs etc. To increase their potential, these and a handful of other such databases have been integrated into a single resource, InterPro, which rationalises their overlaps and adds biological annotation and cross-references to diverse data sources. InterPro is used for automatic annotation of proteins both in the protein databases and within genome annotation projects. This chapter reviews the major protein signature methods and the databases in the public domain that implement the methods, and goes on to describe the InterPro resource.

Keywords: Protein family, signature, clustering, functional classification, hidden Markov model, regular expression, profile.

Introduction

The number of nucleotide and protein sequences in public sequence databases is increasing exponentially as sequencing of whole genomes becomes commonplace. Due to their origin, these sequences are primarily unannotated and require further analysis to functionally characterise them. Sequences have traditionally been classified through sequence similarity searches, which are successful in many ways, however, limited by the database searched, and in their ability to detect distant homologs. With these databases becoming biased towards unannotated sequences, new sequences have a high likelihood of not receiving functional annotation from a sequence similarity search. Therefore, the more characterised sequences in the database, the more efficient sequence similarity searches will be for functional annotation. Limitations in this approach have led to the development of new methods for automatic functional annotation of proteins, and these rely heavily on the concept of protein families.

A protein family is generally described as a group of evolutionarily related proteins, and the basis for classifying proteins into families is their sequence similarity. A protein may be related to another very specifically at the sub-family level, it can be related to more diverse proteins at a family level, and it can be related to even more diverse proteins at the superfamily level. The number of common functional properties of the proteins within each set increases from the superfamily towards the sub-family level. This feature of protein families is used to transfer annotation from well-characterised proteins to related, but uncharacterised protein family members. The possibility of classifying proteins into families thus presents an opportunity for automatic functional annotation of proteins.

There are two main methods for classification of proteins into families: sequence clustering and protein signatures. Methods of clustering related protein sequences are based on sequence similarity searches and have the drawbacks associated with them. As mentioned previously, an important factor is the low probability of detecting distant members of a protein family. The second approach uses protein signatures to classify proteins into families by exploiting known similarities between members of the family and using mathematics to describe them. A signature is a "description" of a protein family or domain, and defines the characteristics associated with the family/domain such that these characteristics can be recognised in new family members. These approaches have different advantages and disadvantages and are used in combination to complement each other. A number of public databases have developed in parallel that use one or more of these methods, and they have been integrated into a single resource known as InterPro [Mulder *et al.*, 2005]. This chapter describes sequence clustering and protein signature methods and the databases that use them, and how they have been integrated into InterPro to provide a useful tool for the automatic annotation of proteins.

Sequence Clustering and Prodom

Sequence clustering methods are generally entirely automated, and make the assumption that members of a protein family will cluster together based on sequence similarity. The methods used for clustering have been described in detail in the previous chapter. Another example of a database that uses sequence clustering methods is ProDom [Servant et al., 2002, Bru et al., 2005]. ProDom uses the concept that autonomously evolving protein domains can be recognized as shuffled in present day sequences by sequence comparison methods. The ProDom method starts with the UniProt protein Knowledgebase (UniProtKB) [Apweiler et al., 2004], determines the smallest sequence in the database, after removing fragments, and searches the remainder of the database with this as the query sequence using PSI-BLAST [Altschul et al., 1997]. All matching sequences are removed and used to create a new ProDom domain family. The remaining sequences are once again sorted by size to identify the smallest, and the process is repeated until all UniProt sequences are classified into their domain families. Due to their methods, ProDom has high coverage of protein sequence space, but also has some very small and potentially biologically meaningless clusters in their 150 000 families.

ProDom provides additional information regarding their protein families in the form of potential gene and protein names, size and consistency of cluster, 3D structure data and taxonomic range information. They calculate the most frequent gene and protein names from the set of proteins in each family automatically, and provide a graphical view of all proteins, showing their domain composition with other ProDom domains. In addition to the main ProDom database, the team offers ProDom-CG and ProDom-SG. ProDom-CG facilitates analysis of ProDom domains in whole genomes. A user can search for domains not represented in a particular taxonomic group, or those found in at least one of the kingdoms archae, bacteria or eukaryotes, or only those domains found in all these kingdoms. ProDom-SG provides the structural genomics candidate search, in which a user can identify all families and related families linked to the PDB, or families and related families with no links to the PDB. These results can then be filtered by taxonomic group, size of family, etc. This facility is to allow structural genomics groups to narrow down the search for potential candidates [Bru et al., 2005].

Protein Signatures and Databases

Sequence clustering methods have the advantage of high coverage, however, automated methods do not take biological information into account. Therefore, alternative methods for classification of proteins into families have been developed. In sequence analysis, a single protein sequence can provide limited information aside from the properties of each amino acid residue. However, if a number of related sequences are available, an alignment can identify conserved residues, which may be important for function. These conserved residues in related proteins provide a means for determining the characteristics of sequences within a family such that these characteristics can be used to identify additional related proteins. There are different methods for describing these characteristics in a computer-readable format, including regular expressions, profiles and Hidden Markov Models (HMMs). The majority of protein signature databases available to the public use one or more of these methods.

Regular Expressions

A regular expression is a formula for matching strings that follow a pattern in some text. In the context of this chapter, regular expressions describe a group of amino acids that constitute a short, conserved motif within a protein sequence, for example a binding site or an active site.

PROSITE [Hulo *et al.*, 2004] is a database that uses such regular expressions, also known as patterns. These are built from sequence alignments of related sequences, which are taken from either a well-characterised protein family derived from the literature, or from the result of sequence searches against UniProtKB/Swiss-Prot and UniProtKB/TrEMBL [Boeckmann *et al.*, 2003]. The alignments are scanned for conserved regions involved in the catalytic activity or substrate binding. Patterns derived from these regions specify which amino acid(s) may or may not occur at each position. The syntax of a PROSITE pattern can be illustrated by the following hypothetical pattern: <M-G-x(3)-[IV]2-x-{FWY}, which is restricted at the N-terminus of a sequence (<) and translated as Met- Gly-any residue-any residue- any residue-[Ile or Val]-[Ile or Val]-any residue-{any residue but Phe or Trp or Tyr}. Patterns have many advantages, but they also have their limitations across whole sequences and for more divergent proteins [Hulo *et al.*, 2004].

Profiles

A profile is a table of position-specific amino acid weights and gap costs. The table describes the probability of finding a particular amino acid at a given position in the sequence [Gribskov *et al.*, 1990]. The numbers in the table (scores) are used to calculate similarity scores between a profile and a sequence for a given alignment. For each set of sequences a threshold score is calculated to differentiate between true and non-true matches to the profile.

PROSITE creates profiles for their database to complement the patterns [Bucher *et al.*, 1996]. For these they also start with multiple sequence alignments, and use a symbol comparison table to convert residue frequency distributions into weights, resulting in a table of position-specific weights [Gribskov *et al.*, 1990]. A similarity score is calculated between the profile and sequences in UniProtKB/Swiss-Prot, and the profile is refined until only the intended set of protein sequences scores above the threshold score for the profile.

PRINTS [Attwood *et al.*, 2003] uses the concept of profiles in a different way to develop their database of "fingerprints". A fingerprint is a group of conserved motifs used to characterize a protein family. Instead of focussing on small conserved areas only, the occurrence of these conserved areas across the whole sequence is taken into account. Profiles are built from a multiple sequence alignment for small conserved regions in the sequence, and together make up a fingerprint. During creation of the fingerprints, each motif is used to scan the protein sequence database and the hit lists are correlated to add sequences to the original alignment. New motifs are then generated and the process is repeated until convergence. Recognition of individual elements in the fingerprint is mutually conditional, and true members match all elements in order while members of a subfamily may match part of the fingerprint. Many fingerprints have been created to identify proteins at the superfamily as well as the family and sub-family levels. For this reason many of the fingerprints are related to each other in an ordered hierarchical structure [Attwood *et al.*, 2003].

Hidden Markov Models

Hidden Markov models [Krogh *et al.*, 1994] are statistical models that are based on probabilities rather than on scores. The HMMER package, written by Sean Eddy (http://hmmer.wustl.edu/), is based on Bayesian statistical models and is the major implementation of HMMs. It allows users to create an HMM from a sequence alignment and to search a database of sequences against the HMM. Many databases, for example Pfam [Bateman *et al*, 2004], SMART [Letunic *et al.*, 2004], TIGRFAMs [Haft *et al.*, 2003], PIRSF [Wu *et al.*, 2004], SUPERFAMILY [Madera *et al.*, 2004], Gene3D [Pearl *et al.*, 2005] and PANTHER [Mi *et al.*, 2005] use HMMs implemented with their own specific applications.

Pfam [Bateman *et al*, 2004] is a large collection of protein families available via the web and in flat file format. There are two sections in the Pfam database, PfamA, which is a set of high quality manually annotated models, and PfamB, which has higher coverage, but is fully automated with no manual intervention. PfamA entries initiate from a seed alignment of the protein sequence region covered by the family or domain, which is trimmed manually to ensure accurate domain boundaries. A full alignment and an HMM are then created from the final alignment and annotation is added. PfamB is created from automatic clustering of the protein sequences in UniProtKB by ProDom with PfamA members removed. Pfam has a high coverage of sequence space and is used extensively for the annotation of genomes. Pfam also provides additional links to protein 3D structure information and protein/domain interaction data where available [Bateman *et al.*, 2004].

SMART (a Simple Modular Architecture Research Tool) [Letunic *et al.*, 2004] is a database that develops HMMs to facilitate the identification and annotation of genetically mobile domains and the analysis of domain architectures. These are built from hand curated multiple sequence alignments of representative family members based on tertiary structures where possible, otherwise those found by PSI-BLAST [Altschul *et al.*, 1997]. The models created are used to search the database for additional members to be included in the sequence alignment. This iterative process is repeated until no further homologues are detected. SMART focuses on models for domains found in signaling, extracellular and chromatin-associated proteins, and provides useful tools for the analysis of protein domain composition [Letunic *et al.*, 2004].

TIGRFAMs [Haft *et al.*, 2003] are a collection of protein families created to assist in sequence annotation, particularly for microbial genomes. The focus is on HMMs that group homologous proteins conserved with respect to function. The models are produced in a similar way to those in Pfam and SMART, but should only hit equivalogs, ie. proteins that have been shown to have the same function. However, where the biology of a protein family does not support the construction of equivalog models, superfamily HMMs are developed and a hierarchy may exist. TIGRFAMs are a useful complement to Pfam where the latter may only have generalised models for a protein family not necessarily restricted to functional equivalents [Haft *et al.*, 2003].

PIRSF [Wu *et al.*, 2004], previously known as PIR SuperFamily, is a protein classification system based on evolutionary relationships between whole proteins. Preliminary PIRSF clusters are computationally defined using both pairwise-based and cluster-based parameters. In each cluster, proteins sharing the same domain composition and similar lengths are manually identified and grouped as curated families. HMMs are created from these families to populate the PIRSF database. A new sequence must be matched by the HMM and satisfy the length restriction to be a member of a family. The focus on classification of whole proteins, not individual domains, allows annotation of specific biological functions of proteins as well as of generic biochemical functions [Wu *et al.*, 2004].

SUPERFAMILY [Madera *et al.*, 2004] is a database of HMMs based on protein structure rather than sequence. The HMMs in the database represent all proteins of known structure, and are based on the SCOP [Andreeva *et al.*, 2004] classification of proteins. Each model corresponds to a SCOP domain and aims to represent an entire superfamily. Many Pfam domains are also based on SCOP domains, however, structural families often differ from sequence-based ones, since similar sequences may have similar structures but the reverse is not always true. SUPERFAMILY has been applied to whole genomes and is used for genome comparisons and analysis of the distribution of structural families in the genomes.

Gene3D [Pearl *et al.*, 2005] is another structure-based set of HMMs, which are built from CATH superfamilies. The CATH database classifies protein structural domains into superfamilies and sequence families and expands these families with related sequences from public databases and complete genomes. They use representatives from each family and the SAM-T technology [Karplus *et al.*, 1998] to build their HMMs. These HMMs have been run over a set of approximately 150 complete genomes, in which they achieved between 40% and 60% coverage [Pearl *et al.*, 2005].

PANTHER [Mi *et al.*, 2005] is a large database of over 30,000 HMMs describing protein families and subfamilies. The subfamilies group functionally related proteins and show the divergence of specific functions from the protein families. Their level of functional specificity is designed for function prediction and annotation of protein families. The PANTHER family and subfamily HMMs cover approximately 90% of mammalian genomes, and the database provides associations between subfamilies and biological pathways, where available [Mi *et al.*, 2005].

Integration of the Databases

Each of the methods for protein family classification described above has their advantages and disadvantages. Patterns are easy to generate and useful for small conserved regions like active sites or binding sites, but fail to provide information about the rest of the sequence, and may reject significant family members. Profiles and HMMs are efficient at identifying divergent members, but HMMs, in particular, are slow to develop and use. PRINTS fingerprints are strong in detecting families on different levels, but are weak in domains and sites. The ideal protein classification system would use all of the databases together. Many of these databases already have interactions and exchange of data between them. For example, some use ProDom families as a starting point for their signatures, and many cross-reference each other where applicable. The resolution was provided by several groups, which made an effort to integrate these databases into InterPro [Mulder *et al.*, 2003], a single coherent protein signature resource. InterPro currently integrates signatures from PROSITE, Pfam, PRINTS, ProDom, SMART, TIGRFAMs, PIRSF, SUPERFAMILY, Gene3D and PANTHER.

InterPro

InterPro is produced by the members of the individual protein signature databases, and the integration of signatures from these databases is manual. In InterPro, the similarities and differences between the contributing member database signatures are rationalized to provide the user with a clear view on how to interpret multiple hits to a protein sequence. All signatures representing the equivalent domain or family are merged into single InterPro entries with annotation describing the domain/family. If a signatures describes a subset of proteins or a region on a protein sequence falling within another signature, then relationships are inserted between the corresponding InterPro entries to represent parent/child (family) or contains/found in (domain composition) relationships. A graphical display present in an entry shows which other InterPro entries it overlaps with. In addition to grouping and describing related signatures, InterPro provides a list of all proteins in UniProtKB matching the entry, and cross-references to protein 3D structure information and specialized databases.

Protein Matches

All protein signatures in InterPro are run against all proteins in UniProtKB using the InterProScan software package [Quevillon *et al.*, 2005], which integrates the algorithms from the member databases and provides the output in a unified format. The protein matches are provided in each InterPro entry in various formats, including an overview, which provides a summary of InterPro entries matched by the protein, a detailed view, which displays matches to each signature in the entries, and a table view which lists the matches with match positions and status of the matches (true, false positive, false negative, etc.). The views include protein 3D structure information, where available, with links to the corresponding SCOP [Andreeva *et al.*, 2004], CATH [Pearl *et al.*, 2005] and MSD (Macromolecular Structure Database) [Golovin *et al.*, 2004] databases. Clicking on the Astex [Hartshorn, 2002] icon displays the structure in three dimensions in the AstexViewer. These links are derived from known structures deposited in the Protein Data Bank [Bourne *et al.*, 2004], however, structural predictions from Swiss-Model [Kopp and Schwede 2004] and ModBase [Pieper *et al.*, 2004] are also displayed, if available and where there is no known structure.

A new feature of the protein match graphical displays is the option to view matches for splice variants of a master sequence in UniProtKB. A link is available within the protein match section of an InterPro entry, which takes the user to a page showing the matches for the master sequence and all possible isoforms in either the overview or detailed view format.

These views provide information on how the isoforms can vary in domain composition and function (Figure 1).

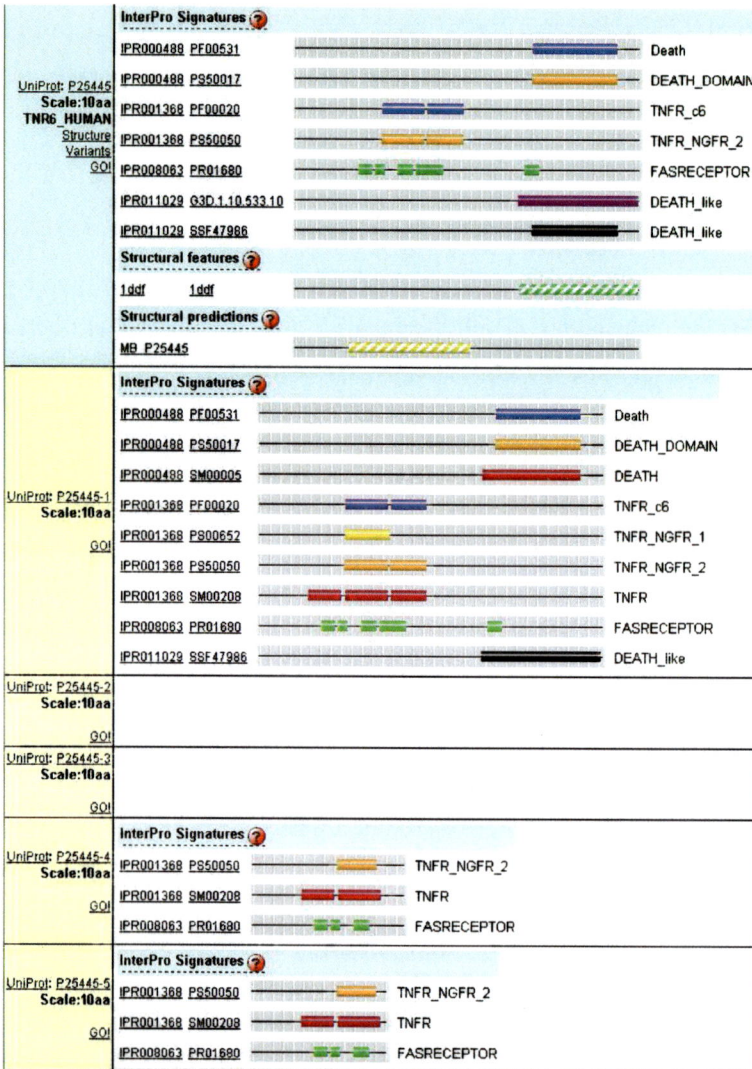

Figure 1. This figure shows an example of the InterPro detailed view of matches for UniProtKB sequence P25445, human calmodulin, and its isoforms. The InterPro matches for the master sequence are in the first block and its isoforms below. The figure shows that isoforms 2 and 3 have sequences that no longer hit InterPro, and the matches for isoforms 4 and 5 vary from the master sequence

Cross-References to other Databases

In addition to cross-references to structural data, InterPro provides links to specialised databases. Where all proteins matching an entry have a conserved function or biological process, the InterPro entry is mapped to Gene Ontology (GO) terms [Harris *et al.*, 2004]. These mappings can be assumed to be applicable to all proteins matching the entry, thus providing a way of large-scale mapping of proteins to GO terms. Currently InterPro-to-GO mappings account for the largest number of protein-to-GO annotations available. Automatic

GO assignment is available in InterProScan, which displays the potential GO terms for query sequences based on the InterPro-to-GO mappings.

InterPro entries provide links to taxonomy databases and show the taxonomic range of proteins matching each entry. It is possible to view the matches for proteins from certain taxonomic groups alone. InterPro also provides links to specialised protein family databases, such as the MEROPS database of proteases and protease inhibitors [Rawlings *et al.*, 2004], the Carbohydrate Active Enzyme (CAZy) database [Coutinho and Henrissat, 1999], and the Enzyme Database (http://www.ebi.ac.uk/intenz/index.html).

InterPro Access

The InterPro database is freely available to the public at http://www.ebi.ac.uk/interpro, which provides access to the database via text searches or browsing the entries by type (family, domain, repeat, active site, binding site or post-translational modification). User documentation is provided, together with help pages for user support. InterProScan is accessible from the web server, an email server or web services for searching query sequences, and a stand-alone version is provided on the InterPro FTP site for downloading and running locally if bulk or confidential searches are required. The InterPro entry and match data is also available for downloading in XML format on the FTP site, together with a MySQL dump of the database.

Conclusion

The classification of proteins into families is the first step in the process of functionally annotating unknown proteins in the database. The methods and databases described in this chapter are crucial to this process and although they have developed independently, they have involuntarily converged to provide a reliable combination for use in the automatic annotation of protein sequences. InterPro unites them to exploit their strengths and compensate for potential weaknesses and provides the bench scientists and genome sequencing centres with a tool for analysis of their protein sequences.

InterPro is a powerful tool for classifying and characterising proteins that is scalable for application to whole genomes of many diverse organisms. The protein signatures in InterPro entries group proteins into families or those with common domains, which are presumed to have conserved functions. Entries representing protein subfamilies tend to contain smaller protein sets, but are more likely to have conserved functions. If one or more proteins matching an entry are well characterised, it is possible to make assumptions about the function of all the other members using rules and automatic annotation algorithms. Having a defined method for grouping the proteins means that new protein sequences entering the database can automatically be assigned to their relevant protein families and thus functionally annotated. This annotation can also be extended to GO terms. InterPro is currently used for the automatic annotation of TrEMBL proteins, and forms an integral part of many automatic genome annotation pipelines.

Protein signatures provide an efficient and reliable alternative to sequence similarity searches for protein function prediction and annotation. They are more sensitive and there are many parameters that can be adjusted to ensure the maximum accuracy. However, there needs to be a balance between sensitivity and specificity, a lack of the former leads to increased numbers of false negatives, while a low specificity results in many false positives. When protein signatures are used in automatic annotation, it is important to achieve reasonable coverage, but without encouraging false predictions. Protein classification methods provide useful and often very accurate predictions of protein function, but these should subsequently be verified by experimental data. Curator-driven annotation of UniProtKB/Swiss-Prot, described later in this book uses a combination of InterPro results, sequence analysis tools and experimental evidence to more accurately describe the protein function.

Acknowledgements

The author would like to acknowledge the InterPro Consortium for their contributions to the success of the InterPro database.

References

Altschul SF, Madden TL, Schaffer AA, Zhang J, Zhang Z, Miller W and Lipman DJ (1997) Gapped BLAST and PSI-BLAST: a new generation of protein database search programs. *Nucleic Acids Research,* 25(17), 3389-3402.

Andreeva A, Howorth D, Brenner SE, Hubbard TJ, Chothia C and Murzin AG (2004) SCOP database in 2004: refinements integrate structure and sequence family. *Nucleic Acids Research,* 32(1), D226-229.

Apweiler R, Bairoch A, Wu CH, Barker WC, Boeckmann B, Ferro S, Gasteiger E, Huang H, Lopez R, Magrane M, Martin MJ, Natale DA, O'Donovan C, Redaschi N and Yeh LS (2004) UniProt: the Universal Protein knowledgebase. *Nucleic Acids Research*, 32(1), D115-119.

Attwood TK, Bradley P, Flower DR, Gaulton A, Maudling N, Mitchell AL, Moulton G, Nordle A, Paine K, Taylor P, Uddin A and Zygouri C (2003) PRINTS and its automatic supplement pre-PRINTS. *Nucleic Acids Research*, 31(1), 400-402.

Bateman A, Coin L, Durbin R, Finn RD, Hollich V, Griffiths-Jones S, Khanna A, Marshall M, Moxon S, Sonnhammer EL, Studholme DJ, Yeats C and Eddy SR (2004) The Pfam protein families database. *Nucleic Acids Research*, 32(1), D138-141.

Boeckmann, B, Bairoch A, Apweiler R, Blatter MC, Estreicher A, Gasteiger E, Martin MJ, Michoud K, O'Donovan C, Phan I, Pilbout S and Schneider M (2003) The SWISS-PROT protein knowledgebase and its supplement TrEMBL in 2003. *Nucleic Acids Research,* 31, 365-370.

Bourne, P. E., Westbrook, J. and Berman, H. M. (2004) The Protein Data Bank and lessons in data management. *Briefings in Bioinformatics,* 5(1), 23-30.

Bru C, Courcelle E, Carrere S, Beausse Y, Dalmar S and Kahn D (2005). The ProDom database of protein domain families: more emphasis on 3D. *Nucleic Acids Research,* 33(1), D212-215.

Bucher P, Karplus K, Moeri N and Hofmann K (1996) A flexible motif search technique based on generalized profiles. *Computational Chemistry,* 20(1), 3-23.

Coutinho, P.M. and Henrissat, B. (1999) Carbohydrate-Active Enzymes server at URL: http://afmb.cnrs-mrs.fr/CAZY/

Golovin A, Oldfield TJ, Tate JG, Velankar S, Barton GJ, Boutselakis H, Dimitropoulos D, Fillon J, Hussain A, Ionides JM, John M, Keller PA, Krissinel E, McNeil P, Naim A, Newman R, Pajon A, Pineda J, Rachedi A, Copeland J, Sitnov A, Sobhany S, Suarez-Uruena A, Swaminathan GJ, Tagari M, Tromm S, Vranken W and Henrick K. (2004) E-MSD: an integrated data resource for bioinformatics. *Nucleic Acids Research,* 32, D211-216.

Gribskov M, Luthy R and Eisenberg D (1990) Profile analysis. *Methods in Enzymology,* 183, 146-159.

Haft DH, Selengut JD and White O (2003) The TIGRFAMs database of protein families. *Nucleic Acids Research,* 31, 371-373.

Harris MA, Clark J, Ireland A, Lomax J, Ashburner M, Foulger R, Eilbeck K, Lewis S, Marshall B, Mungall C, Richter J, Rubin GM, Blake JA, Bult C, Dolan M, Drabkin H, Eppig JT, Hill DP, Ni L, Ringwald M, Balakrishnan R, Cherry JM, Christie KR, Costanzo MC, Dwight SS, Engel S, Fisk DG, Hirschman JE, Hong EL, Nash RS, Sethuraman A, Theesfeld CL, Botstein D, Dolinski K, Feierbach B, Berardini T, Mundodi S, Rhee SY, Apweiler R, Barrell D, Camon E, Dimmer E, Lee V, Chisholm R, Gaudet P, Kibbe W, Kishore R, Schwarz EM, Sternberg P, Gwinn M, Hannick L, Wortman J, Berriman M, Wood V, de la Cruz N, Tonellato P, Jaiswal P, Seigfried T, White R and Gene Ontology Consortium. (2004) The Gene Ontology (GO) database and informatics resource. *Nucleic Acids Research,* 32(1), D258-261.

Hartshorn MJ (2002) AstexViewer: a visualisation aid for structure-based drug design. *Journal of Compututationally Aided Molecular Design,* 16(12), 871-881.

Hulo N, Sigrist CJ, Le Saux V, Langendijk-Genevaux PS, Bordoli L, Gattiker A, De Castro E, Bucher P and Bairoch A (2004) Recent improvements to the PROSITE database. *Nucleic Acids Research,* 32(1), 134-137.

Karplus K, Barrett C and Hughey R. (1998) Hidden Markov models for detecting remote protein homologies. *Bioinformatics,* 14, 846–856.

Kopp J and Schwede T. (2004) The SWISS-MODEL Repository of annotated three-dimensional protein structure homology models. *Nucleic Acids Research,* 32(1), D230-234.

Krogh A, Brown M, Mian IS, Sjolander K and Haussler D (1994) Hidden Markov models in computational biology. Applications to protein modeling. *Journal of Molecular Biology,* 235(5), 1501-1531.

Letunic I, Copley RR, Schmidt S, Ciccarelli FD, Doerks T, Schultz J, Ponting CP and Bork P (2004) SMART 4.0: towards genomic data integration. *Nucleic Acids Research,* 32(1), D142-144.

Madera M, Vogel C, Kummerfeld SK, Chothia C and Gough J (2004) The SUPERFAMILY database in 2004: additions and improvements. *Nucleic Acids Research*, 32(1),D235-239.

Mi H, Lazareva-Ulitsky B, Loo R, Kejariwal A, Vandergriff J, Rabkin S, Guo N, Muruganujan A, Doremieux O, Campbell MJ, Kitano H and Thomas PD. (2005) The PANTHER database of protein families, subfamilies, functions and pathways. *Nucleic Acids Research,* 32(1), D284-288.

Mulder, N. J., Apweiler, R., Attwood, T. K., Bairoch, A., Bateman, A., Binns, D., Bradley, P., Bork, P., Bucher, P., Cerutti, L., Copley, R., Courcelle, E., Das, U., Durbin, R., Fleischmann, W., Gough, J., Haft, D., Harte, N., Hulo, N., Kahn, D., Kanapin, A., Krestyaninova, M., Lonsdale, D., Lopez, R., Letunic, I., Madera, M., Maslen, J., McDowall, J., Mitchell, A., Nikolskaya, A. N., Orchard, S., Pagni, M., Ponting, C. P., Quevillon, E., Selengut, J., Sigrist, C. J., Silventoinen, V., Studholme, D. J., Vaughan, R. and Wu, C. H. (2005) InterPro, progress and status in 2005. *Nucleic Acids Research,* 33(1), D201-5.

Pearl F, Todd A, Sillitoe I, Dibley M, Redfern O, Lewis T, Bennett C, Marsden R, Grant A, Lee D, Akpor A, Maibaum M, Harrison A, Dallman T, Reeves G, Diboun I, Addou S, Lise S, Johnston C, Sillero A, Thornton J and Orengo C. (2005) The CATH Domain Structure Database and related resources Gene3D and DHS provide comprehensive domain family information for genome analysis. *Nucleic Acids Research,* 33(1), D247-51.

Pieper U, Eswar N, Braberg H, Madhusudhan MS, Davis FP, Stuart AC, Mirkovic N, Rossi A, Marti-Renom MA, Fiser A, Webb B, Greenblatt D, Huang CC, Ferrin TE and Sali A. (2004) MODBASE, a database of annotated comparative protein structure models, and associated resources. *Nucleic Acids Research,* 32, D217-222.

Quevillon E, Silventoinen V, Pillai S, Harte N, Mulder N, Apweiler R and Lopez R. (2005) InterProScan: protein domains identifier. Nucleic Acids Research, 33, W116-20.

Rawlings ND, Tolle DP and Barrett AJ. (2004) MEROPS: the peptidase database. *Nucleic Acids Research*, 32, D160-164.

Servant F, Bru C, Carrere S, Courcelle E, Gouzy J, Peyruc D and Kahn D (2002) ProDom: automated clustering of homologous domains. *Briefings in Bioinformatics,* 3(3), 246-251.

Wu CH, Nikolskaya A, Huang H, Yeh LS, Natale DA, Vinayaka CR, Hu ZZ, Mazumder R, Kumar S, Kourtesis P, Ledley RS, Suzek BE, Arminski L, Chen Y, Zhang J, Cardenas JL, Chung S, Castro-Alvear J, Dinkov G and Barker WC (2004) PIRSF: family classification system at the Protein Information Resource. *Nucleic Acids Research*, 32(1), D112-114.

In: In Silico Genomics and Proteomics
Editors: N. Mulder and R. Apweiler, pp. 37-54

ISBN 1-59454-995-8
© 2006 Nova Science Publishers, Inc.

Chapter 4

The Gene Ontology Annotation (GOA) Database: Sharing Biological Knowledge with GO

Evelyn Camon, Daniel Barrell, Emily Dimmer and Vivian Lee
European Bioinformatics Institute, Wellcome Trust Genome Campus, Hinxton,
Cambridge, UK, CB10 1SD

Abstract

Gene Ontology (GO) is an established dynamic and structured vocabulary that has been successfully used in protein annotation. Designed by biologists to improve data integration, GO attempts to replace the multiple nomenclatures used by specialised and large knowledgebases. This chapter describes the Gene Ontology Annotation (GOA) database which uses the GO vocabulary to provide high quality electronic and manual annotations to the proteins contained in UniProt Knowledgebase. As a supplementary archive of GO annotation, GOA promotes a high level of integration of the knowledge represented in UniProt with other databases.

1. Introduction

To realize the full potential of the human proteome, the research community needs to access and analyse the most up-to-date sequence and knowledge from all eukaryotes. This research is hindered when users have to query across many isolated datasets, each with their own nomenclature. Clearly, keeping pace with the increasing volume of data and user demands requires continual improvement and optimization in how data is organised and shared. This is an ongoing challenge for the world's biological databases. Recently, it has become apparent that successful integration is reliant on each database using the same language. In this regard, one of the most important advances in data interoperability has been the cooperative development of semantic standards, not only in computing but also addressing the biological content of our databases.

In 1998 the Gene Ontology Consortium (Harris *et al*, 2004) pioneered a community effort to develop a set of structured vocabularies and standards for the data representation of *molecular function*, *biological process* and *cellular component*. These GO vocabularies have now grown to nearly 20,000 terms and carry definitions to indicate their intended usage (personal communication Midori Harris, April 2005). GO terms are arranged in a directed acyclic graph (DAG) wherein any term may have more than one parent as well as zero, one or more children. This structure makes attempts to describe biology much richer than would be possible with a hierarchical graph. As GO is both flexible and dynamic, the terms and their location in the DAG are maintained by a team of dedicated GO editors. Tracking changes to GO is also important, so each GO term is identified by an unique stable GO ID (e.g. GO:0005554). Although GO is primarily designed by biologists for biologists, it has also been used in ways for which it was not originally designed, for example the development of information extraction technologies (Camon *et al*, 2005). It is unlikely that GO will ever achieve perfection or completeness, however it can claim to be widely used by the biological community as it has citations in over 700 publications. The success of GO is largely based on its open source approach and the involvement throughout its development of various biological communities rich in expertise and committed to its upkeep. GO's success has also been propelled by its early uptake by a number of key biological databases including the UniProt Knowledgebase (UniProtKB),(Bairoch *et al*, 2005).

As the world's most highly annotated protein sequence database, UniProtKB has archived and annotated nearly two million proteins through a combination of manual and electronic techniques. By 2010, it is estimated this figure will increase to over six million proteins, the majority of which will lack biochemical and functional characterization. One way in which to maximise the annotation of this data while safeguarding quality, is to draw on the expertise of in-house and specialist community resources. In this respect, the UniProtKB has always been at the forefront, by cross-referencing its records with over 88 different specialised databases. The creation of GO made it possible to go one step further and directly integrate the curated biological knowledge. In 2001, the UniProt group became a member of the GO Consortium, thus demonstrating its support for the GO initiative. It initiated the Gene Ontology Annotation (GOA) project (Camon *et al*, 2003, Camon *et al*, 2004) to provide a dedicated database and curation team for the assignment of GO terms to all well-characterised proteins and in particular, to that of the human proteome.

Five years on, the initial aims and objectives of the GOA project have been achieved. GOA has organized, shared and integrated protein knowledge using the GO structured vocabulary. GOA is now one of the largest contributors to the GO annotation effort, with over 9 million GO annotations to nearly two million proteins, covering over 100,000 species (as of March 2006). This chapter will detail both the electronic and manual GO annotation processes and describe how these data are distributed and used by biological and bioinformatics communities.

2. GO Annotation Process

High quality GO annotations (GOA) are generated through a combination of electronic and manual techniques, the latter of which employs a team of skilled biologists (See GOA Data flow Figure 1). GOA recommends that users wishing to analyse GO annotation understand how GO is arranged and how GO assignments are made. Guidelines for GO annotation have been detailed before and are published on the GO home page (Table 2(p)).

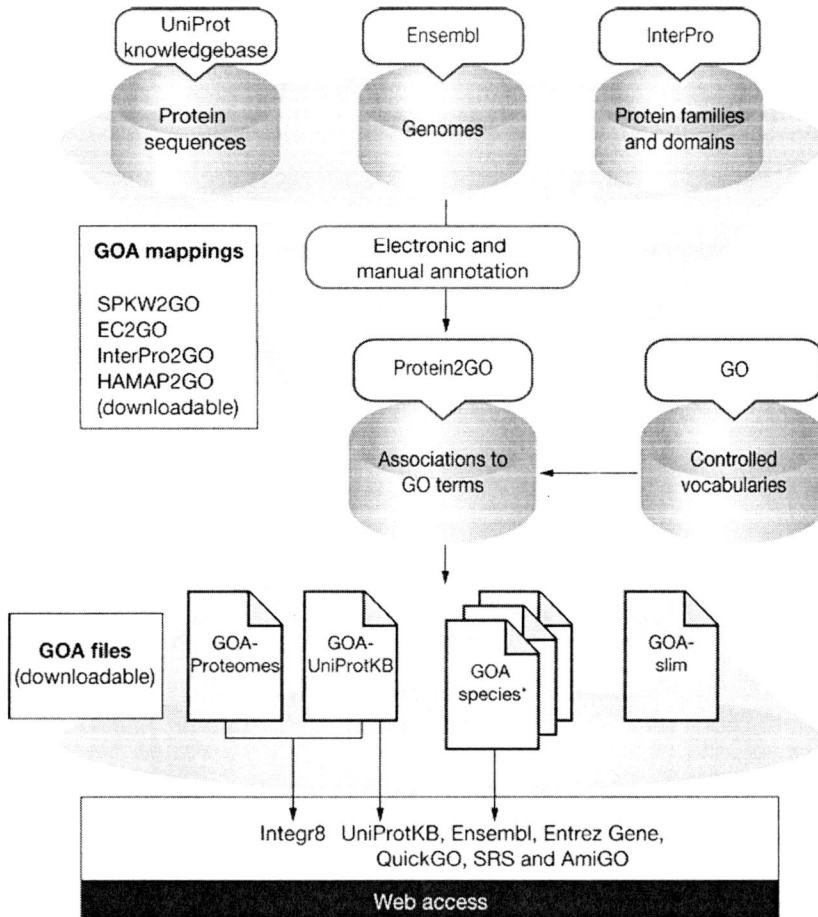

*Nonredundant data sets for human, mouse, rat, cow, chicken, *Arabidopsis* and zebrafish

Figure 1. GOA dataflow. Manual and electronic protein to GO annotations from in-house and external resources are integrated in our GOA database. Tab-delimited files of associations for a selection of species as well as for all species in UniProtKB are available for download. Mappings files and annotations to GO-slim files are also updated and released regularly.

2.1. Large Scale GO Annotation Using Electronic Techniques

The first phase of the GOA project involved the large-scale assignment of GO terms to UniProtKB entries using electronic methods. This strategy involved converting a proportion of the pre-existing knowledge held within the flat files into GO terms (Camon *et al*, 2003). For example, UniProt description lines [DE] may contain Enzyme Commission (EC) numbers. By using an existing manual mapping of EC numbers to the GO molecular function ontology , EC2GO (e.g. EC:1.1.1.1 > GO:alcohol dehydrogenase activity ; GO:0004022) and a mapping of UniProt accession numbers to EC numbers, GOA can produce a protein to GO association. "Mapping" here refers to the linking of various classification systems to GO terms, while the word "association" refers to a connection between a database object (gene, transcript, or protein) and a GO term.

The curation of UniProtKB records includes the assignment of keywords. These serve as a subject reference for each sequence and assist in the retrieval of specific categories of data. Keywords can be assigned manually after careful curation of the literature and sequence analysis checks. They can also be added to UniProtKB entries during automatic annotation of the UniProtKB/TrEMBL database (Apweiler *et al*, 2001). The latter procedure utilizes a novel system of standardized transfer of annotation from well-characterized proteins in UniProtKB/Swiss-Prot to unannotated entries with similar sequence or structure (Kretschmann *et al,* 2001). Currently, UniProtKB maintains a controlled list of 895 keywords, each with a definition to clarify its biological meaning and intended usage. As of March 2005, 68% of these terms have been manually mapped to the GO vocabulary. Keywords not mapped include those that have multiple usages, have no equivalent GO term, or are beyond the scope of the GO project, such as keywords describing diseases. The Swiss-Prot keyword to GO (SPKW2GO) mapping file is routinely used to generate a large number of annotations to the GO process, function and component ontologies. The accuracy of the association of GO terms to UniProtKB entries is based on the keywords within them and is assured by the annotation quality standards already existing in UniProtKB.

The use of bi-directional database cross-references also helps to integrate GO annotations. For example, the majority of UniProtKB entries will cross-reference an InterPro identification number (e.g. IPR004367) and vice versa. InterPro is a key database maintained at the EBI (Mulder *et al,* 2005). It provides an integrated documentation resource for proteins, families and domains. A single InterPro entry provides comprehensive annotation describing a set of related proteins, some of which may have identical functions, be involved in the same processes, and act in the same locations. During the curation of each InterPro entry, high-level GO terms are manually mapped after inspecting all available information. In each case, the abstracts and the annotation of proteins within the match lists are read and an appropriate GO term mapped if it applies to the whole protein. Some entries can be mapped to very deep level (granular) GO terms, while entries describing wider families or common domains can only be mapped to higher level terms or can not be mapped at all. The associated GO term therefore applies to all proteins with true hits to all signatures in the InterPro entry. An advantage of this technique is that multifunctional proteins can be mapped to multiple GO terms through associations with more than one InterPro entry. So far, the application of the InterPro2GO mapping in the electronic assignment of GO terms to proteins has produced the most coverage in the GOA dataset. In a similar fashion, UniProtKB is supplemented with GO annotations originating from a High-quality Automated and Manual Annotation of microbial Proteomes (HAMAP) (Gattiker *et al*, 2003) to GO mapping, a manually curated set of GO terms to microbial protein families.

SPKW2GO, InterPro2GO and HAMAP2GO mappings are maintained by UniProt and distributed on the GO and EBI FTP sites on a regular basis. To further support interoperability, InterPro2GO has been used to generate GO mappings to the member databases of InterPro and these also are available for download (e.g. PFAM2GO). The integrity of the mappings is maintained by running regular sanity checks on the data. These checks include searching for mappings from secondary or deleted accession numbers and mappings to obsolete GO terms. The reports are manually corrected.

The resultant GO associations are released monthly, in accordance with a GO Consortium agreed format, within an 'association file' (a 15 column tab-delimited file). As the mapping files used by GOA are manually curated, GOA is confident that its electronic annotation is of a high standard. In fact, a recent evaluation for the text mining community verified that these techniques could precisely predict the correct GO term in 94-100% of cases with 72% recall (Camon *et al*, 2005). Despite this high performance, it is still important for users to have the ability to distinguish electronic from manually curated data. For these reasons, the certainty of each GOA association is supported by annotating to one of 12 consortium-agreed evidence codes (Table 1). Electronically generated associations are labelled with the evidence tag 'inferred from electronic annotation' or IEA to alert users that these annotations have not been manually checked for each protein entry, and therefore only 'likely' to be involved in a particular GO activity. These IEA assignments are popular amongst the GO Consortium and external research community in generating first-pass annotations to a variety of species.

Table 1. GO Consortium evidence codes and example usage

Code	Evidence Full Name	An Example Use
IEA	Inferred Electronic Annotation	Annotation based on mapping e.g. InterPro2GO
TAS	Traceable Author statement	Annotation based on author statement from e.g. review paper
NAS	Non-traceable Author Statement	Annotation based on direct author statement from experimental paper
IC	Inferred from Curator judgement	Annotation based on curator knowledge and expertise
IDA	Inferred from Direct Assay	Annotation based on direct assay e.g. immunofluorescence
IPI	Inferred from Physical Interaction	Annotation based on e.g. 2-hybrid interaction
IGI	Inferred from Genetic Interaction	Annotation based on genetic assay e.g. functional complementation
IEP	Inferred from Expression Pattern	Annotation based on transcript/protein levels e.g. microarray/Western Blot.
IMP	Inferred from Mutant Phenotype	Annotation based on e.g. a knock-out experiment
ISS	Inferred from Sequence or structural Similarity	Annotation based on sequence alignment analysis by curator or author
RCA	Inferred from Reviewed Computational Analysis	Annotation based on a non-sequence-based computational method, where the results have been reviewed by an author /curator.
ND	No biological Data available	Used for annotations to "unknown" molecular function, biological process, or cellular component terms when no supporting evidence available.

2.2. High Quality Manual GO Annotation

The large-scale assignment of GO terms to UniProtKB proteins using electronic methods is a fast and efficient way of associating high-level terms to a large number of proteins. However, to provide more reliable and specific annotation, the GOA project also makes use of manual curation using information extracted from published scientific literature (Camon *et al*, 2004). This process is slower than the use of electronic techniques but provides more accurate information, as all annotation is validated by a team of skilled biologists. Each manually assigned term is associated with a GO experimental evidence code (Table 1, Table 2(r)) and a PubMed ID. These enable users to track the exact literature source and type of experiment used to support the annotation (Figure 2A).

Priority is given in the GOA project to the annotation of data from the human proteome. This complements the efforts of the other consortium members as no other member is responsible for providing human-specific data. The GOA project also assigns GO terms to proteins from a wide range of other species. New terms are requested as required to adequately describe the functional aspects of proteins from many species in UniProtKB, thus enhancing the GO ontologies and extending their scope.

Each GO consortium member uses slightly different techniques for manual GO annotation (Hill *et al*, 2004, Dwight *et al*, 2002). This variation depends on the size and expertise of curation staff, the use of an organism for experimental analysis and the size of the genome requiring annotation. Curation of organisms with rich experimental data focuses primarily on extracting GO annotation from the literature, while the curation of organisms with less experimental data tends to use more comparative inferences for GO annotation (personal communication Judy Blake). The following describes the annotation strategy of the GOA curators.

Three categories of human proteins are prioritized for GO annotation, (a) those which have no GO annotation, (b) those which have disease relevance and (c) those which are important for microarray analyses. Having chosen the protein accession to annotate, the relevant scientific papers need to be found. The first step is to decide if the papers already linked within the UniProt entry are relevant for GO annotation. The decision on whether to read the full text of a paper is based on the curator's interpretation of the text used in the paper title or abstract. The journals cited in UniProtKB/TrEMBL records are inherited from EMBL/DDBJ/GenBank databases (Kanz *et al*, 2005) and so may describe the sequence rather than GO function, process or component. Papers that reference the sequence are accompanied by a remark located in the reference position (RP) line, which says 'NUCLEOTIDE SEQUENCE'. On the other hand, UniProtKB/Swiss-Prot records are manually supplemented with documents to support the annotation stored in the comment (CC) lines. In these cases, the remark in the RP line might also indicate the type of information extracted from a paper e.g. 'FUNCTION', 'SUBCELLULAR LOCATION' or 'INTERACTION'. It should be noted, however, that the use of the word FUNCTION in Swiss-Prot is not the same as 'Molecular Function' usage in GO. Frequently, GO process terms can be extracted from FUNCTION CC lines. Further reading on the scope of each GO category can be found on the GO home page (Table 2 s-u).

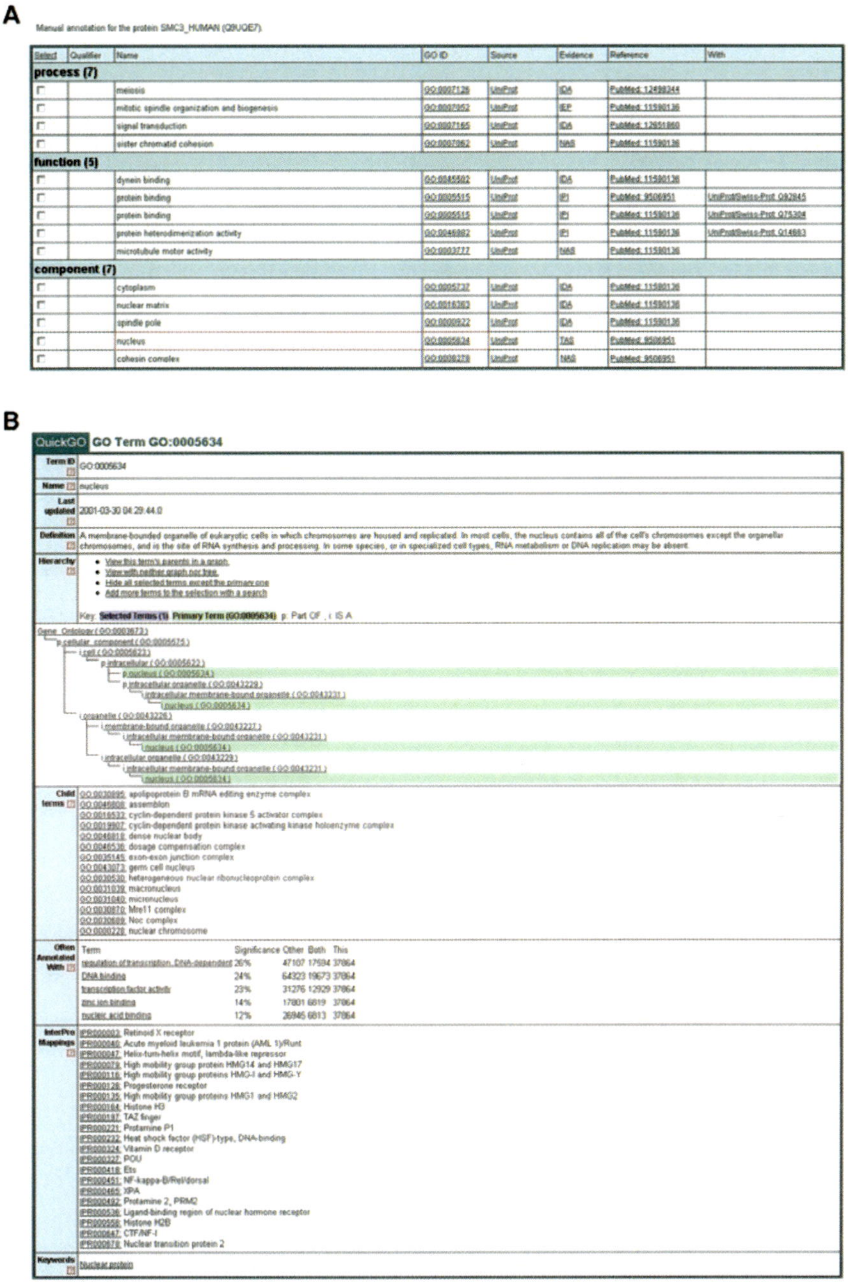

Figure 2. (A) Sample of QuickGO display page showing all GO terms that have been manually annotated to the protein SMC3_HUMAN (UniProt ID Q9UQE7). Each term is associated with a GO evidence code (e.g. IDA) and a PubMed ID to allow users to track and hyperlink to the literature and type of evidence (see Table 1) used to support the annotation. For annotations which are inferred from physical interactions (IPI), the UniProt accession for protein which is bound is indicated in the 'with' column. (B) Sample of the QuickGO GO term page for "nucleus" using the denormalized tree view for its parents. Information about child GO terms, concurrent annotations and its mappings to InterPro or Swiss-Prot keywords are also shown. This figure has been edited to fit this page

In addition to the papers archived in the UniProtKB records, the NCBI PubMed advanced search (Wheeler *et al*, 2005) or EBI's CiteExplore (http://www.ebi.ac.uk/citations) are queried to find papers that support supplementary GO annotation. Various combinations of the gene and protein, full and abbreviated names are searched. Initially, searches are limited to 'title' or 'title/abstract' and to 'human entries only'. Electronic GO annotation and information in UniProtKB/Swiss-Prot CC lines often provide curators with an insight into the types of functions that could be extracted from the literature. With this information to hand, curators are able to refine their search options to find more than enough relevant papers for GO annotation. In GOA, our current aim is to find the most recent papers which provide experimental evidence for the unique features of a given protein. Our approach is protein-centric rather than paper-centric as it is not yet necessary or possible to read all the relevant papers that might be used to assign the same GO term. In the future, however, adding more papers to experimentally verify a given function will provide greater confidence to the GO annotations. A good source of a complete set of functional annotations is often retrieved from recent review articles. These reports often have links to relevant papers with experimental verification. Any papers that report new data are fed back to the UniProt curators to add to the original entry.

Most GO Consortium members would agree that the most difficult task in searching the literature is finding papers that have experimental information for a given species. Often, the species 'name' (e.g. human) is not mentioned in the title or abstract and occasionally, not directly mentioned in the full text. On these occasions, the method section of the paper has to be read and perhaps the taxonomic origin of a cell line identified before any attempt at GO curation. Filtering 'human entries only' via PubMed is not always accurate. In addition, authors do not always cite the most up-to-date gene nomenclature e.g. use of upper case letters for human gene symbols (Wain *et al*, 2004). This is also likely to affect the precision of automatic gene product entity extraction techniques.

2.3. Locating Functional Annotation and Choosing the Correct GO Term

Once a relevant paper is found, the full text is read to identify the unique features of a given protein. The majority of papers will mention more than one protein; however a curator will concentrate on capturing the information pertinent to the main protein chosen for annotation. Currently most curators still prefer to print out papers rather than view papers online. This is simply to limit computer eye strain and because a curator can quickly scan and select the most relevant parts of the document for curation. Words or short phrases which can be converted to GO terms are highlighted by hand and the correct GO term identifier (ID) is documented in the paper margins for review. In the future this curation step is likely to be assisted by the emerging information extraction technologies.

GO terms are chosen by querying the GO ontologies with the QuickGO (Table 2(a), Figure 2B) or AmiGO (Table 2(b)) web browsers or with a local copy of OBO-Edit, the official Open Biomedical Ontology editor which also has browsing capability (Table 2(y)). Before assigning a GO term, the definition must be read to check its suitability. Obsolete GO terms are not used in annotation (Table 2(v)). There are currently over 1000 GO terms tagged as obsolete within the GO database. It is important that users involved in text mining remove these GO terms from ontology downloads before performing any extraction analysis (Camon *et al*, 2005). When electronic or manual GO annotations become obsolete, they are manually replaced with an appropriate term. The reason for the obsoletion and suggestions for replacement GO terms are documented in GO comment lines. If a useful term is missing from the ontology, an existing GO term is in the incorrect hierarchical position or a definition needs to be refined, a curator request is sent to the GO editorial office using an online tracker system hosted by SourceForge (Table 2(x)).

The GO Consortium avoids using species-specific definitions for GO terms; however some functions, processes and components are not common to all organisms. Inappropriate species-specific GO terms (e.g. germination GO:0009844) should not be manually annotated to mammalian proteins. Sometimes these species-specific terms can be distinguished by the *sensu* (in the sense of) designation (e.g. embryonic development (sensu Magnoliophyta), GO:0009793). Curators are cautious when manually assigning these terms. To avoid generating inappropriate GO term assignments, the text mining community should read the GO Consortium documentation on the subject (Table 2(w)).

If a curator is unsure of which process term should accompany a function term, they can consult the 'Often Annotated With' section of the QuickGO browser. Here, GO terms that are often assigned in tandem are displayed. These are also referred to as common concurrent assignments and are calculated on our existing manual GO annotations (Camon *et al*, 2003). For example the GO function term 'nucleus' (GO:0005634) (Figure 2B) is often co-assigned with the GO component term 'DNA binding' (GO:0003677).

It is important to note that GO terms are often extracted from particular regions of a paper. The 'Materials and Methods' section of a paper is only used to identify the species of protein used in the research and to determine which GO experimental evidence code should be used. It is not used to extract functional annotation. Furthermore, curators often piece together information from different parts of a document to reinforce a decision to annotate. GO annotation is not associated with a UniProt protein until the entire article is read.

If no functional annotation can be found for a given protein after an exhaustive literature search, the GO root terms molecular_function (GO:0003674), biological_process (GO:0008150) or cellular_component (GO:0005575) can be assigned with GO evidence code ND ('No Data'). This evidence code when provided together with an annotation date indicates to the next curator that a recent attempt to characterise this protein was unsuccessful. After a reasonable time lapse, a second attempt would be made.

3. Data Searching and Retrieval

There are various ways of accessing and searching GOA project data (Table 2). In addition to several web-based browsers, GOA files and mappings can be downloaded (Camon *et al,* 2004). GOA is also cross-referenced in numerous GO Consortium databases as well as Ensembl (Hubbard *et al*, 2005), Entrez Gene (Maglott *et al*, 2005), Reactome (Joshi-Tope *et al*, 2005) and EMBL/GenBank/DDBJ databases.

**Table 2. Summary of sites for browsing or
downloading the GOA dataset and essential reading**

	Resource	Description
	Web-based tools	
a	QuickGO	http://www.ebi.ac.uk/ego
b	AmiGO	http://www.godatabase.org
c	UniProtKB	http://www.ebi.uniprot.org
d	Integr8	http://www.ebi.ac.uk/integr8
e	InterPro	http://www.ebi.ac.uk/interpro
f	SRS	http://srs.ebi.ac.uk
	Downloads	
h	Association Files	ftp://ftp.ebi.ac.uk/pub/databases/GO/goa
i	Xref Files	ftp://ftp.ebi.ac.uk/pub/databases/GO/goa
j	GOA-Proteomes	http://www.ebi.ac.uk/GOA/proteomes
k	Annotations to GOA GO-Slim	ftp://ftp.ebi.ac.uk/pub/databases/GO/goa/goslim
l	Mappings to GO	ftp://ftp.ebi.ac.uk/pub/databases/GO/goa/external2go
m	BioCreative Results	http://www.pdg.cnb.uam.es/BioLINK/workshop_BioCreative_04/results
	Essential Reading	
n	GOA Home Page	http://www.ebi.ac.uk/GOA
o	GO Tools	http://www.geneontology.org/GO.tools
p	GO Annotation Guide	http://www.geneontology.org/GO.annotation
q	GO Bibliography	http://www.geneontology.org/GO.biblio
r	GO Evidence codes	http://www.geneontology.org/GO.evidence
s	GO Function	http://www.geneontology.org/GO.function.guidelines
t	GO Process	http://www.geneontology.org/GO.process.guidelines
u	GO Component	http://www.geneontology.org/GO.component.guidelines
v	GO Obsolete terms	http://www.geneontology.org/GO.usage.html#obsoleteTerms
w	GO Sensu Definition	http://www.geneontology.org/GO.usage.html#sensu
x	SourceForge	http://sourceforge.net/projects/geneontology http://www.geneontology.org/GO.sourceforge.links
y	OBO	http://obo.sourceforge.net/
z	BioCreative Task 2	http://www.pdg.cnb.uam.es/BioLink/BioCreative_task2

3.1. GOA Association File Release

The most common form of data transfer within the GO Consortium is a tab-delimited file of associations between proteins and GO terms. Currently the GOA group produces six association files (Figure 2,Table 2). GOA-UniProt is the largest, including GO annotations to UniProtKB proteins from over 100,000 species (March 2006). This can be searched via SRS or QuickGO/AmiGO browsers. The GOA group also releases a number of species-specific association files that provide GO annotation to a number of non-redundant proteome sets using data from either the Integr8 (http:www.ebi.ac.uk/integr8) or International Protein Index (IPI) (http://www.ebi.ac.uk/IPI) resources. So far, association files have been created for human, mouse, rat, zebrafish, cow and *Arabidopsis* proteomes. These are built using the IPI resource, which provides maximal, non-redundant protein sets built from the UniProt, Ensembl, TAIR (Rhee *et al*, 2003), RefSeq (Pruitt *et al*, 2005) and H-Invitational databases (http://www.h-invitational.jp/).

The interconversion of identifiers from different systems and databases is a common problem in functional genomics. Therefore as an extra resource, GOA provides conversion tables for each of the species-specific association files to a range of identifiers in other systems (ftp://ftp.ebi.ac.uk/pub/databases/GO/goa/), including the EMBL/Genbank/DDBJ nucleotide sequence databases, HUGO, and Entrez Gene and RefSeq at NCBI. There are also a variety tools available for the automatic mapping of identifiers between systems during expression analysis e.g. MatchMiner (Bussey *et al*, 2003).

3.2. GOA-Proteomes

As well as the species-specific files (produced in conjunction with the IPI dataset), GOA also provides monthly releases of a GO association files for a selection of completely sequenced proteomes. These files are built only from UniProtKB accessions (ftp://ftp.ebi.ac.uk/pub/databases/GO/goa/proteomes) and provide GO annotations to proteins from species whose proteome is in the public domain and which has more than 25% GO coverage. For these files the UniProt non-redundant proteome set is provided by the Integr8 project (Table 2 (d)).

4. Utility of GO Annotation

The success of the GO Consortium can already be measured by the number of world-leading academic and commercial databases that use it to annotate and exchange biological knowledge in a consistent manner. As further proof of its popularity, a query to PubMed (March 2005) for the phrase 'gene ontology' will return over 700 articles. This is a considerable achievement since 80% of all papers are cited just 10 times or less (Redner, 1998). The articles which cite GO or GOA describe many different uses of the ontologies and the annotation files. The following section will describe the primary uses of GOA.

4.1. Use of GO Annotation in Database Querying and Navigation

Biologists expect to navigate and query proteomic data repositories in a biologically intuitive and scientifically accurate manner. The UniProtKB provides a wide variety of features and keywords to assist this process. GO annotation offers an additional approach. For example, users querying the UniProtKB by protein/gene name or disease relevance would also retrieve the supplementary GO annotation. Human entries which have no experimental proof of function may be supplemented with GO annotation from mouse or rat orthologs. In this situation the experimental source (mouse UniProt accession) of the GO annotation is always displayed in the human entry along with the GO evidence code 'inferred from sequence similarity' (Table 1).

As GO represents a universal set of curated keywords, users can also retrieve all possible annotations to a high level GO terms in a candidate-based approach. This is because according to GO philosophy, every child term inherits the meaning of all of their parent terms. As such, every annotation to a child term should be true for every parent of that child; therefore known as the 'true path rule'. Some GO queries are quite broad, for example finding all proteins involved in the biological process of 'transcription'. This query would involve the retrieval of all proteins not only annotated to the GO term for 'transcription' (GO:0006350) but also to the proteins assigned to the children of this GO term e.g. transcription, DNA-dependent (GO:0006351). Users can also directly query the database for proteins involved in more specific GO processes such as 'snoRNA transcription' (GO:0009302). Retrieving the annotations to the children and parent GO term is currently possible via AmiGO and SRS (Table 2 (b)(f)).

In the future, cross-referencing with other Open Biomedical Ontologies offers the possibility of even more complicated biological queries. For example, to identify "which proteins have G-coupled protein receptor activity and have been implicated in depression"? would involve querying the database with a combination of terms from both the GO and a disease ontology. The ability to answer specific biological problems can only improve with time as more databases increase their usage of ontologies to share annotation. For the moment GO annotation in most databases is incomplete and so an absence of protein to GO annotation does not necessarily mean a lack of function.

4.2. Use of GO in Gene Expression Data Analysis

The increasing use of gene expression profiling offers great promise for clinical research into disease biology and its treatment. Along with the ability to measure changing expression levels in thousands of genes at once, comes the challenge of analyzing and interpreting the vast sets of data generated. There are numerous analysis tools under development to meet these challenges, many of which try to interpret the observed expression changes in terms of the biological properties of the genes being analysed (Doniger *et al*, 2003; Volinia *et al*, 2004). GO annotation provides such a link between biological knowledge and the genes studied in the expression profiles. When combined with statistical analysis, the GOA dataset is a useful resource for building pathways and can help facilitate the design of more focused experiments for validating the relevance and importance of these processes in human disease and therapeutics. For example, in a recent study (Seo *et al*, 2005) GOA helped to expedite the molecular characterisation of proteins up-regulated in cervical cancer. Not only were several cancer-specific cellular processes identified, genes of no known function were also highlighted for further analysis. Whether these GO processes can be used to predict prognosis and diagnosis of cancer patients however, remains to be seen. An example of GO annotation usage in the clustering of gene expression data is shown in Figure 3.

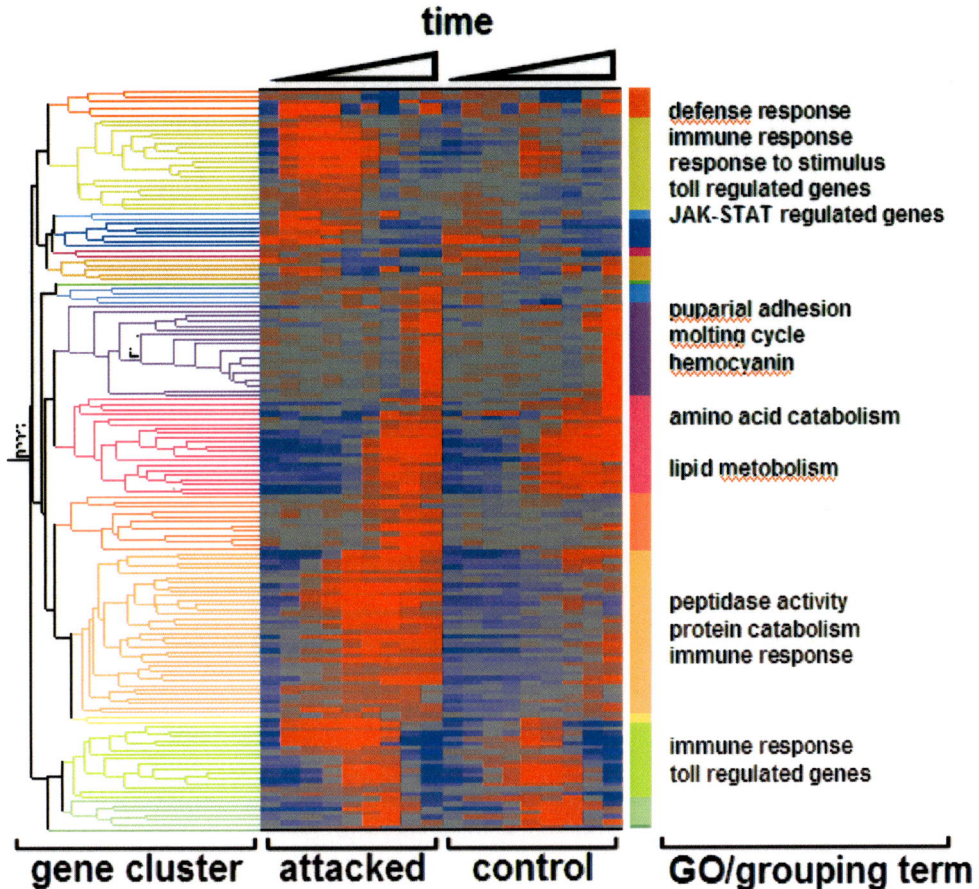

Figure 3. Use of GO annotation in the clustering a whole genome microarrays. This figure shows changes in gene expression in response to parasitoid attack in *Drosophila* larvae (Wertheim *et al*, 2005)

In addition to microarray analysis, GOA is also incorporated into evolutionary studies, particularly when correlating structure to function relationships (Shakhnovich et al, 2003). By including GO annotation directly into mass spectrometry results, users can quickly understand the biological significance of the identified proteins (Blonder et al, 2004). For further reading on the variety of GO tools available consult the GO Consortium tool page (Table 2(o)).

4.3. Use Annotations to GO Slim Terms to Simplify Display and Analysis

With the continual increase in range and depth of coverage of GO, users can find it difficult to gain an overview of the activities or subcellular locations to which a set of proteins have been annotated.

Slimmed down versions of the GO can provide users with such a perspective. GO-slims are set of high-level and wide-ranging terms, which provides users with an overview of the GO and its accompanying annotations.

GO slims have proved to be particularly useful for giving a summary of the GO annotations of a proteome, microarray, or cDNA collection when broad classification of gene product function is required (Figure 3).

GOA provides a generic GO slim (ftp://ftp.ebi.ac.uk/pub/databases/GO/goa/goslim/), which contains approximately 80 high-level GO terms that provide an overview of the main functions, processes and subcellular localizations of a proteome (Table 2(h)). This GOA slim has been integrated into the GO flat file, so that updates are synchronised with any changes that are applied to the GO. As an additional service GOA also provides a file containing the GOA annotations already mapped to this GOA slim on the EBI FTP site (Table 2(k)). From there, users can, for example, download all existing annotations to the GO term for 'nucleus' (GO:0005634).

As each community has different needs, a variety of GO slim files have been archived on the GO home page by Consortium members (ftp://ftp.geneontology.org/pub/go/GO_slims/). GO slims can also be designed to be specific to species or to particular areas of the ontologies. Customised GO slims can be created using the OBO editor (https://sourceforge.net/project/showfiles.php?group_id=36855) and a map2slim Perl script (http://search.cpan.org/~cmungall/go-perl-0.01/scripts/map2slim) to map annotations up to their parent GO slim terms.

4.4. Use of GO in Validating Text Mining and Information Extraction Techniques

As GOA produces a high quality set of GO annotations, it allows comparisons to be made with new annotation approaches and is an important tool for validation of these methods.

Manual annotation, although reliable, is time consuming and dependent on skilled biologists capable of extracting key information from the published literature. To expedite the GO curation process, the bioinformatics community has focused on the development of new automatic annotation techniques, such as automated information extraction and the conversion of this knowledge into the GO vocabulary (Chiang and Yu, 2003; Muller *et al*, 2004). This has resulted in a variety of GO prediction servers with varying abilities to interpret accurately the subtleties of the scientific natural language as well as GO structure, mappings and annotation styles. To assess these information extraction techniques and allow users to apply the methods judiciously, the BioLINK group (http://www.pdg.cnb.uam.es/ BioLINK/) organised the first BioCreative (Critical Assessment of Information Extraction systems in Biology) competition in winter 2003. In collaboration with BioLINK, GOA provided one of the gold standard training and test sets of GO annotation. UniProt curators also took part in the manual verification of GO terms mined from the literature and corroborated by evidence from the text. In spring 2004, the results of competition were announced. The techniques used in the competition were not ready to supercede current electronic strategies in GOA (Camon *et al*, 2005) in terms of recall and precision. However, this is an emerging research field and GOA contributed significantly to its further development by creating a training set of protein to GO associations linked to the correct text strings.

5. How to Contribute to GOA

The GOA team is pleased to freely provide such a large and valuable resource of GO annotations. The success and accuracy of GOA relies on the support of the biological communities who use and share GO annotation and on our commitment to keep electronic and manual datasets up-to-date. GOA actively encourages all users to improve this resource, by informing us via email (goa@ebi.ac.uk) if a particular GO annotation requires updating.

GOA especially encourages specialist groups also interested in fast-tracking the annotation of the human proteome to initiate collaborations with us, that we might cross-reference and share annotations and provide scientists as complete a resource as possible.

Acknowledgements

The GOA project is grateful for the continued financial support of the NIH grant, 1R01HGO2273-01. GOA is also grateful the past support of the European Commission grants QRLT-2001-00015 and QLRI-2000-00981. Thanks to all colleagues worldwide who have contributed to the information content of the GOA dataset and to the development of annotation and retrieval tools. Special thanks is due also to Jane Lomax for corrections to this chapter and to Eugene Schuster for providing Figure 3.

References

Apweiler R. (2001) Functional information in Swiss-Prot: The basis for large-scale characterisation of protein sequences. *Brief. Bioinform.*, 2(1), 9-18.

Bairoch A., Apweiler R., Wu C.H., Barker W.C., Boeckmann B., Ferro S., Gasteiger E., Huang H., Lopez R., Magrane M., Martin M.J., Natale D.A., O'Donovan C., Redaschi N., and Yeh L.S.. (2005) The Universal Protein Resource (UniProt). *Nucleic Acids Res.*, 1(33), D154–159.

Bussey K.J., Kane D., Sunshine M., Narasimhan S., Nishizuka S., Reinhold W.C., Zeeberg B., Ajay W., Weinstein J.N.,(2003) MatchMiner: a tool for batch navigation among gene and gene product identifiers. *Genome Biol.* 4(4):R27. Epub 2003

Blonder J., Terunuma A., Conrads T.P., Chan K.C., Yee C., Lucas D.A., Schaefer C.F., Yu L.R., Issaq H.J., Veenstra T.D., and Vogel J.C (2004) A proteomic characterization of the plasma membrane of human epidermis by high-throughput mass spectrometry. *J Invest Dermatol.* 123(4), 691-9.

Camon E., Magrane M., Barrell D., Binns D., Fleischmann W., Kersey P., Mulder N., Oinn T., Maslen J., Cox A., and Apweiler R. (2003) The Gene Ontology Annotation (GOA)

Database: sharing knowledge in Uniprot with Gene Ontology. *Genome Res.*, 13(4), 662–672.

Camon E., Magrane M., Barrell D., Lee V., Dimmer E., Maslen J., Binns D., Harte N., Lopez R., and Apweiler R. (2004) The Gene Ontology Annotation (GOA) Database: sharing knowledge in Uniprot with Gene Ontology. *Nucleic Acids Res.*, 1(32), D262–266.

Camon E.B., Barrell D.G., Dimmer E.C., Lee V., Magrane M., Maslen J., Binns D., and Apweiler R. (2005) An evaluation of GO annotation retrieval for BioCreAtIvE and GOA. *BMC Bioinformatics*, 6(Suppl 1). *In Press*

Chiang J.H., and Yu H.C. (2003) MeKE: discovering the functions of gene products from biomedical literature via sentence alignment. *Bioinformatics* 19(11),1417-22.

Doniger S.W., Salomonis N., Dahlquist K.D., Vranizan K., Lawlor S.C., and Conklin B.R. (2003) MAPPFinder: using Gene Ontology and GenMAPP to create a global gene-expression profile from microarray data. *Genome Biol.* 4(1):R7. Epub 2003.

Dwight S.S., Harris M.A., Dolinski K., Ball C.A., Binkley G., Christie K.R., Fisk D.G., Issel-Tarver L., Schroeder M., Sherlock G., Sethuraman A., Weng S., Botstein D., and Cherry J.M. (2002) Saccharomyces Genome Database (SGD) provides secondary gene annotation using the Gene Ontology (GO).*Nucleic Acids Res* 2002, 30(1):69-72.

Gattiker A, Michoud K, Rivoire C, Auchincloss AH, Coudert E, Lima T, Kersey P, Pagni M, Sigrist CJ, Lachaize C, Veuthey AL, Gasteiger E, and Bairoch A. (2003) Automated annotation of microbial proteomes in SWISS-PROT. *Comput Biol Chem.* 27(1), 49-58.

Harris M.A., Clark J., Ireland A., Lomax J., Ashburner M., Foulger R., Eilbeck K., Lewis S., Marshall B., Mungall C., Richter J., Rubin G.M., Blake J.A., Bult C., Dolan M., Drabkin H., Eppig J.T., Hill D.P., Ni L., Ringwald M., Balakrishnan R., Cherry J.M., Christie K.R., Costanzo M.C., Dwight S.S., Engel S., Fisk D.G., Hirschman J.E., Hong E.L., Nash R.S., Sethuraman A., Theesfeld C.L., Botstein D., Dolinski K., Feierbach B., Berardini T., Mundodi S., Rhee S.Y., Apweiler R., Barrell D., Camon E., Dimmer E., Lee V., Chisholm R., Gaudet P., Kibbe W., Kishore R., Schwarz E.M., Sternberg P., Gwinn M., Hannick L., Wortman J., Berriman M., Wood V., de la Cruz N., Tonellato P., Jaiswal P., Seigfried T., and White R.; Gene Ontology Consortium. (2004) The Gene Ontology (GO) database and informatics resource. *Nucleic Acids Res.*, 1(32), D258–261.

Hill D.P., Begley D.A., Finger J.H., Hayamizu T.F., McCright IJ., Smith C.M., Beal J.S., Corbani L.E., Blake J.A., Eppig J.T., Kadin J.A., Richardson J.E., and Ringwald M. (2004) The mouse Gene Expression Database (GXD): updates and enhancements. *Nucleic Acids Res*, 1(32), D568-571.

Hubbard T., Andrews D., Caccamo M., Cameron G., Chen Y., Clamp M., Clarke L., Coates G., Cox T., Cunningham F., Curwen V., Cutts T., Down T., Durbin R., Fernandez-Suarez X.M., Gilbert J., Hammond M., Herrero J., Hotz H., Howe K., Iyer V., Jekosch K., Kahari A., Kasprzyk A., Keefe D., Keenan S., Kokocinsci F., London D., Longden I., McVicker G., Melsopp C., Meidl P., Potter S., Proctor G., Rae M., Rios D., Schuster M., Searle S., Severin J., Slater G., Smedley D., Smith J., Spooner W., Stabenau A., Stalker J., Storey R., Trevanion S., Ureta-Vidal A., Vogel J., White S., Woodwark C., and Birney E. (2005) Ensembl 2005. *Nucleic Acids Res*, 1(32), D568-571.

Joshi-Tope G., Gillespie M., Vastrik I., D'Eustachio P., Schmidt E., de Bono B., Jassal B., Gopinath G.R., Wu G.R., Matthews L., Lewis S., Birney E., and Stein L. Reactome: a knowledgebase of biological pathways. *Nucleic Acids Res*, 1(33), D438-432.

Kanz C., Aldebert P., Althorpe N., Baker W., Baldwin A., Bates K., Browne P., van den Broek A., Castro M., Cochrane G., Duggan K., Eberhardt R., Faruque N., Gamble J., Diez F.G., Harte N., Kulikova T., Lin Q., Lombard V., Lopez R., Mancuso R., McHale M., Nardone F., Silventoinen V., Sobhany S., Stoehr P., Tuli M.A., Tzouvara K., Vaughan R., Wu D., Zhu W., and Apweiler R. (2005) The EMBL Nucleotide Sequence Database. *Nucleic Acids Res*, 1(33), D29-33.

Kretschmann E., Fleischmann W., and Apweiler R. (2001) Automatic rule generation for protein annotation with the C4.5 data mining algorithm applied on Swiss-Prot. *Bioinformatics* (17), 920-926.

Maglott D, Ostell J, Pruitt KD, and Tatusova T. (2005) Entrez Gene: gene-centered information at NCBI. *Nucleic Acids Res*, 1(33), D54-58.

Mulder N.J., Apweiler R., Attwood T.K., Bairoch A., Bateman A., Binns D., Bradley P., Bork P., Bucher P., Cerutti L., Copley R., Courcelle E., Das U., Durbin R., Fleischmann W., Gough J., Haft D., Harte N., Hulo N., Kahn D., Kanapin A., Krestyaninova M., Lonsdale D., Lopez R., Letunic I., Madera M., Maslen J., McDowall J., Mitchell A., Nikolskaya A.N., Orchard S., Pagni M., Ponting C.P., Quevillon E., Selengut J., Sigrist C.J., Silventoinen V., Studholme D.J., Vaughan R., and Wu C.H. (2005) InterPro, progress and status in 2005. *Nucleic Acids Res.*, 1(33), D201–205.

Muller H.M., Kenny E.E., and Sternberg P.W. (2004) Textpresso: an ontology-based information retrieval and extraction system for biological literature. *PLoS Biol.* 2(11):e309. Epub 2004.

Pruitt K.D., Tatusova T., and Maglott D.R. (2005) NCBI Reference Sequence (RefSeq): a curated non-redundant sequence database of genomes, transcripts and proteins. *Nucleic Acids Res*, 1(33), D501-504.

Redner S. (1998) How Popular is your paper? An Emperical study of citation distribution. *Eur.Phys.J.* B4, 131-134.

Rhee S.Y., Beavis W., Berardini T.Z., Chen G., Dixon D., Doyle A., Garcia-Hernandez M., Huala E., Lander G., Montoya M., Miller N., Mueller L.A., Mundodi S., Reiser L., Tacklind J., Weems D.C., Wu Y., Xu I., Yoo D., Yoon J., and Zhang P. (2003) The Arabidopsis Information Resource (TAIR): a model organism database providing a centralized, curated gateway to Arabidopsis biology, research materials and community. *Nucleic Acids Res.* 31(1),224-8.

Seo M.J., Bae S.M., Kim Y.W., Kim Y.W., Hur S.Y., Ro D.Y., Lee J.M., Namkoong S.E., Kim C.K., and Ahn W.S. (2005) New approaches to pathogenic gene function discovery with human squamous cell cervical carcinoma by gene ontology. *Gynecol Oncol.* 96(3):621-629.

Shakhnovich, B.E., Dokholyan, N.V., DeLisi, C. and Shakhnovich, E.I. (2003) Functional fingerprints of folds: evidence for correlated structure-function evolution. *J Mol Biol.*, 326, 1-9.

Volinia S., Evangelisti R., Francioso F., Arcelli D., Carella M., and Gasparini P. (2004) GOAL: automated Gene Ontology analysis of expression profiles. *Nucleic Acids Res.*, 1(32), W492–499.

Wain H.M., Lush M.J., Ducluzeau F., Khodiyar V.K., and Povey S. (2004) Genew: the Human Gene Nomenclature Database, 2004 updates. *Nucleic Acids Res.*, 1(32), D255–257.

Wertheim B., Kraaijeveld A.R., Schuster E., Blanc E., Hopkins M., Pletcher S.D., Strand M.R., Partridge L., and Godfray H.C. (2005) Genome-Wide Gene Expression in Response to Parasitoid Attack in Drosophila. *PLOS Biology.,*6(11):R94.

Wheeler D.L., Barrett T., Benson D.A., Bryant S.H., Canese K., Church D.M., DiCuccio M., Edgar R., Federhen S., Helmberg W., Kenton D.L., Khovayko O., Lipman D.J., Madden T.L., Maglott D.R., Ostell J., Pontius J.U., Pruitt K.D., Schuler G.D., Schriml L.M., Sequeira E., Sherry S.T., Sirotkin K., Starchenko G., Suzek T.O., Tatusov R., Tatusova T.A., Wagner L., and Yaschenko E. (2005) Database resources of the National Center for Biotechnology Information. *Nucleic Acids Res* , 1(33), D39-45.

In: In Silico Genomics and Proteomics
Editors: N. Mulder and R. Apweiler, pp. 55-70

ISBN 1-59454-995-8
© 2006 Nova Science Publishers, Inc.

Chapter 5

Curator-Driven Protein Sequence Analysis and Annotation

Michele Magrane and Claire O'Donovan

EMBL Outstation - European Bioinformatics Institute, Wellcome Trust Genome Campus,
Hinxton, Cambridge CB10 1SD, UK

Abstract

While automated annotation methods provide a rapid means of large-scale protein analysis and characterisation, manual annotation is a slower process which involves literature-based curation and rigorous sequence analysis. Despite its slower pace, it is essential in supplying high-quality information to database users and in providing the accurate data necessary for the development of many automated methods. This chapter details the UniProt Knowledgebase manual curation system and includes a description of the sequence analysis methods used during this process.

1. Introduction

The number of completely sequenced genomes continues to increase, providing huge amounts of data to the scientific community, and efforts are now being focused on the identification and functional analysis of the proteins encoded by these genomes. Automated annotation methods can do much to enhance the information available about uncharacterised proteins as these methods provide a fast and effective means of large-scale analysis and characterisation. Several automatic methods exist, including high-level sequence similarity searches against characterised proteins, and more complex methods which collect the results of alignment and prediction tools to provide automatic characterisation of proteins, such as Pedant (Riley et al., 2005) and GeneQuiz (Hoersch et al., 2000). Another approach (Apweiler et al., 2001) groups well-characterised proteins in the Swiss-Prot section of the UniProt Knowledgebase (UniProtKB/Swiss-Prot) (Bairoch et al., 2005) using InterPro (Mulder et al.,

2005) and transfers the shared annotation to uncharacterised proteins in the TrEMBL section (UniProtKB/TrEMBL).

All of these methods provide a rapid automated means of analysis and increase the amount of information available about uncharacterised proteins. However, they all have in common the fact that they rely on the availability of accurately annotated proteins to enable them to make predictions. Manual annotation is thus essential in providing high-quality information to database users and in supplying the accurate information which acts as the basis of automated methods, and protein sequence databases play a vital role as central resources for storing these data and making them freely available to the scientific community.

Manual annotation is a slow labour-intensive process and is generally the rate-limiting step in the production of any curated biological database. Curation of protein sequences involves enriching basic sequence data with additional information from a range of sources such as scientific literature and sequence analysis tools. This requires identification and prioritisation of relevant characterisation papers from the huge pool of available publications, extraction and validation of information from these papers, and manual confirmation of the results generated by the analysis programs. The use of sequence analysis tools also requires the constant evaluation and comparison of current prediction methods to ensure that the most appropriate and accurate programs are used so that database users are offered the most valid results. The advantage of the manual curation approach is that all information added during the curation process is verified by expert biologists and the data in manually curated collections can be considered to be highly reliable.

The UniProt Knowledgebase (UniProtKB) is a curated protein sequence database which provides a high level of annotation, minimal redundancy and a high level of integration with other biological databases. This chapter details its manual curation process, including a description of the sequence analysis tools used.

2. Capturing the Correct Sequence

For many genes, multiple sequence reports from different groups exist in the collaborative nucleotide sequence databases of DDBJ (Tateno et al., 2005), EMBL (Kanz et al., 2005) and GenBank (Benson et al., 2005). These databases are archive collections where each report for a given gene is stored in its own database entry and this policy differs from that of the UniProt Knowledgebase which is non-redundant. Translations from nucleotide database entries are added to UniProtKB/TrEMBL, the preliminary computer-annotated section of the Knowledgebase, and some automatic redundancy removal is carried out at this stage. A conservative approach is adopted to prevent merging of sequences which may not originate from the same gene so that only sequences from the same species which are 100% identical are merged automatically. This means that, even after this process, there is still some redundancy in the database so the first step in the annotation process is to find all sequence reports which correspond to a particular protein and to merge these into a single entry containing the most correct sequence. This provides users with a highly integrated view of the information available about a particular protein and is often the most complicated aspect of the curation process.

2.1. Sequence Similarity Searches

To identify potential merge candidates, curators perform similarity searches. These searches also identify similar entries which are already fully curated and from which standardised annotation may be transferred and this helps to ensure consistent annotation across protein families. A number of tools can be used for this purpose, depending on the type and length of the sequence in question, as each method provides a unique environment. The main methods used are MPsrch (Sturrock and Collins, 1993), BLAST (Altschul et al., 1997) and FASTA (Pearson and Lipman, 1988). MPsrch is a sequence comparison tool which implements the true Smith and Waterman algorithm. It is the most sensitive method available and is very reliable for distantly related members. It also reports fewer false negatives than other methods (Harte et al., 2004). FASTA can be very specific when identifying long regions of low similarity while BLAST is the fastest similarity search program (Harte et al., 2004).

2.2. Sequence Comparison

Once all sequences have been identified, the next step is to merge them into a single entry. Potential merge candidates are checked thoroughly to ensure that they represent the same protein by comparing gene names and chromosomal location. Sequence comparison is then performed using a variety of multiple sequence alignment methods to identify the exact differences between the sequences. A wide range of methods exist and these have been compared in detail elsewhere (Notredame, 2002). The two most widely used methods are ClustalW (Thompson et al., 1994) and T_Coffee (Notredame et al., 2000). A recently developed method, MUSCLE (Edgar, 2004), is also used as it works well with large numbers of sequences.

These methods allow identification of sequence discrepancies and help in identifying the underlying causes. Polymorphisms and disease mutations may account for some differences while differences affecting large regions, including large gaps and insertions, may be due to alternative splicing events. While some differences are due to naturally occurring events, others arise as a result of sequencing errors. The quality of nucleotide sequencing has improved due to the development of sophisticated automated sequencing methods but there is a still a large amount of heterogeneity in the sequence quality provided by different groups. Sequence comparisons can therefore help in determining the consensus sequence at positions which vary between groups. There is currently an average of two sequence reports in each UniProtKB/Swiss-Prot entry and this number increases for organisms which are the focus of multiple sequencing efforts. This redundancy helps in the identification of potential sequencing errors such as frameshifts and can also help to identify incorrect protein predictions such as selection of an incorrect N-terminal initiation site or incorrect prediction of exon/intron boundaries in genomic sequences. As well as comparison with multiple sequence reports from the same organism, it is also useful to perform sequence comparisons between orthologous and paralogous proteins from related species to give a fuller picture of what the correct sequence may be. Detailed examination of the various sequences available for a particular protein is a very time-consuming process but it ensures that the sequence presented for each protein is as complete and correct as possible and increases the accuracy of the results produced from further sequence analysis.

3. Sequence Analysis

The area of sequence analysis has developed significantly in recent years and the collection of prediction tools currently available can provide valuable information about an uncharacterised sequence. This section provides an overview of the sequence analysis tools used by UniProtKB curators. Most of these tools have been incorporated into the annotation platform and provide an essential starting point for manual annotation. They cover a range of predictions in three main areas: identifying features of a sequence, predicting post-translational modifications and predicting subcellular location. The prediction methods described below are listed in Table 1 along with their URLs.

3.1. Identifying Features of a Sequence

3.1.1. Patterns and Profiles
Databases of signatures diagnostic for protein families, domains or functional sites are important tools for the functional classification of newly determined sequences that lack biochemical characterisation. During the last decade, several signature recognition and sequence clustering methods have evolved to address different sequence analysis problems. The automatic analysis of a protein sequence is possible through the use of protein signatures, which are methods for diagnosing a domain or characteristic region of a protein family. A number of protein signature databases exist, each using a variation on the signature methods available, which include patterns, profiles and hidden Markov models (HMMs). These databases are most effective when used together rather than in isolation, as the strengths and weaknesses of each method are complemented and a comprehensive overview is provided. InterPro (Mulder et al., 2005) integrates the major protein signature databases into one resource: PROSITE (Hulo et al., 2005), which uses regular expressions and profiles, PRINTS (Atwood et al., 2003), which uses position-specific scoring matrix-based fingerprints, ProDom (Bru et al., 2005), which uses automatic sequence clustering, and Gene3D (Pearl et al., 2005), PANTHER (Mi et al., 2005), Pfam (Bateman et al., 2004), PIRSF (Wu et al, 2004), SMART (Letunic et al., 2004), SUPERFAMILY (Gough et al., 2001) and TIGRFAMs (Haft et al., 2003), all of which use HMMs. The coverage of protein space by InterPro continues to grow and has increased to 78% for UniProtKB. Since UniProtKB curators combine these automatic predictions with literature-sourced experimental data, it has been possible to establish a mutually beneficial feedback loop from UniProtKB to InterPro with regard to false positive, false negative and partial matches. This enables the continuing improvement of the methods and prediction results available to curators.

Table 1. List of prediction methods and their URLs

Resource	Method	URL
Similarity searches	MPsrch	http://www.ebi.ac.uk/MPsrch/
	BLAST	http://www.ebi.ac.uk/blast2/index.html
	FASTA	http://www.ebi.ac.uk/fasta/
Sequence	ClustalW	http://www.ebi.ac.uk/clustalw/
comparison	T_Coffee	http://igs-server.cnrs-mrs.fr/~cnotred/Projects_home_page/ t_coffee_home_page.html
	Muscle	http://phylogenomics.berkeley.edu/cgi-bin/muscle/ input_muscle.py
Protein signature	InterPro	http://www.ebi.ac.uk/interpro/
resources	PROSITE	http://www.expasy.org/prosite/
	Pfam	http://www.sanger.ac.uk/Software/Pfam/
	PRINTS	http://umber.sbs.man.ac.uk/dbbrowser/PRINTS/
	ProDom	http://prodes.toulouse.inra.fr/prodom/current/html/home.php
	SMART	http://smart.embl-heidelberg.de/
	TIGRFAMs	http://www.tigr.org/TIGRFAMs/index.shtml
	PIRSF	http://pir.georgetown.edu/iproclass/
	Superfamily	http://supfam.mrc-lmb.cam.ac.uk/SUPERFAMILY/
	Gene3D	http://www.biochem.ucl.ac.uk/bsm/cath/Gene3D/
	PANTHER	https://panther.appliedbiosystems.com/
Transmembrane domains	TMHMM	http://www.cbs.dtu.dk/services/TMHMM/
Coiled-coils	COILS	http://www.ch.embnet.org/software/COILS_form.html
Repeats	REP	http://www.embl-heidelberg.de/~andrade/papers/rep/ search.html
Signal sequences	SignalP	http://www.cbs.dtu.dk/services/SignalP/
Glycosylation	NetNGlyc	http://www.cbs.dtu.dk/services/NetNGlyc/
	NetOGlyc	http://www.cbs.dtu.dk/services/NetOGlyc/
	YinOYang	http://www.cbs.dtu.dk/services/YinOYang/
GPI-anchor	big-PI	http://mendel.imp.univie.ac.at/gpi/cgi-bin/gpi_pred.cgi
	DGPI	http://129.194.185.165/dgpi/index_en.html
Myristoylation	NMT	http://mendel.imp.univie.ac.at/myristate/SUPLpredictor.htm
	Myristoylator	http://www.expasy.org/tools/myristoylator/
Sulfation	Sulfinator	http://www.expasy.org/tools/myristoylator/
Phosphorylation	NetPhos	http://www.cbs.dtu.dk/services/NetPhos/
Subcellular location	TargetP	http://www.cbs.dtu.dk/services/TargetP/
	Predotar	http://genoplante-info.infobiogen.fr/predotar/
	PSORT	http://psort.nibb.ac.jp/

3.1.2 Transmembrane Domains

A variety of tools are available to predict the topology of transmembrane proteins. Such predictions are possible because of the distinctive patterns of hydrophobic (intramembraneous) and polar (loop) regions within a sequence. The percentage of transmembrane proteins does not appear to vary significantly between organisms and about a quarter of all proteins in UniProtKB are predicted to be transmembraneous (Moeller et al., 2001). Membrane proteins play important roles as key components of cell-cell signalling mechanisms, initiating signalling cascades. They also mediate membrane transport of many ions and solutes, as well as being involved in the organism's recognition of self.

Thorough structural analysis of membrane proteins is difficult to achieve due to the intrinsic difficulties involved in growing crystals of membrane proteins. It takes considerably less effort to biochemically determine membrane topology which includes determination of the localisation of the membrane spanning regions and the polarity of their integration into the membrane. Reliable computational methods for topology prediction are very valuable as they provide a basis for further experimental analysis. UniProtKB curators use five distinct transmembrane prediction programs in combination and evaluate the results based on the individual strengths of each program and the conclusions from the combination of results. The programs used are TMHMM (Sonnhammer et al., 1998), ESKW (Eisenberg et al., 1984), RA (Argos and Rao, 1986), KKD (Klein et al., 1985) and MEMSAT (Jones et al., 1984). TMHMM is generally considered to be the best program for predicting the correct number of transmembrane domains (Moeller et al., 2001) while ESKW has proved very accurate in predicting domain lengths.

3.1.3. Coiled-Coils

The super-helical structure called the coiled-coil is formed by the interlocking of two or more alpha-helices which tend to be unstable in water. Examples of proteins with coiled-coil regions include myosins, tropomyosins and intermediate filaments. UniProtKB curators use an adapted version of the COILS program (Lupas et al., 1991) to detect these regions.

3.1.4 Amino Acid-Rich Domains

Each of the 20 amino acids found in proteins can be distinguished by the R-group substitution on the alpha-carbon atom. There are two broad classes of amino acids based upon whether the R-group is hydrophobic or hydrophilic. Hydrophobic amino acids repel the aqueous environment and therefore reside predominantly in the interior of proteins. This class of amino acids does not ionise or participate in the formation of H-bonds. Hydrophilic amino acids tend to interact with the aqueous environment, are often involved in the formation of H-bonds and are predominantly found on the exterior surfaces of proteins or in the reactive centres of enzymes. Therefore, identifying domains rich in particular amino acids within a protein sequence can infer function or location. A program developed in-house is used to detect runs of identical residues. This program looks for at least four identical residues in a row and tries to extend the region in both directions. It will stop if it encounters two residues in a row that are not similar to the core region. A number of PROSITE profiles are also available for detection of sequence regions enriched in a particular amino acid and these are described at http://www.expasy.org/cgi-bin/get-prodoc-entry?PDOC50099.

3.1.5 Repeats

Many proteins possess more than one structural and functional unit, and short protein repeats, frequently with a length of between 20 and 40 residues, represent a significant fraction of known proteins (Andrade et al., 2000). Detection of these repeated units can be problematic due to their sequence divergence, short length and variable number. The REP program (Andrade et al., 2000) is used to detect the following repeats: ankyrin, armadillo, HAT, HEAT, Kelch, leucine-rich repeats, PFTA, PFTB, RCC1, TPR and WD40.

3.2. Post-Translation Modifications

Post-translational modifications are defined as the series of chemical reactions whereby a newly synthesized polypeptide chain is converted to a functional protein. In the post-genome era, it is now accepted that these post-translational modifications are responsible for the great diversity in protein function. They are essential for signal transduction, nuclear import, targeting to membranes, protein folding, and cofactor binding or synthesis. Because of the value of such predictions, this is currently an area of intensive research. However, many prediction methods are in their infancy and have not yet been developed into usable algorithms. The post-translational modification prediction tools most commonly used by UniProtKB curators are for the following modifications: signal peptide cleavage, glycosylation, GPI-anchor modification, N-terminal myristoylation, tyrosine sulfation and phosphorylation.

3.2.1. Signal Peptide Cleavage

Since one of the crucial properties of a protein is its subcellular location, prediction of protein sorting is an important question in bioinformatics. A fundamental distinction in protein sorting is between secretory and non-secretory proteins, determined by a cleavable N-terminal sorting sequence, the secretory signal peptide. SignalP (Neilsen et al., 1997) has been established as the most reliable prediction method (Menne et al., 2000) and is the method of choice for UniProtKB curators. It predicts the presence and location of secretory signal peptide cleavage sites from Gram-positive prokaryotes, Gram-negative prokaryotes and eukaryotes. The method incorporates cleavage site and signal peptide/non-signal peptide predictions, based on a combination of several artificial neural networks and hidden Markov models.

3.2.2. Glycosylation

Glycosylation is an important post-translational modification and is known to influence protein folding, localisation, trafficking, solubility, antigenicity, biological activity and half-life, as well as cell-cell interactions. PROSITE pattern PS00001is used to predict sites of N-glycosylation in extracellular domains for eukaryotic proteins. N-glycosylation of human proteins can be predicted by the NetNGlyc program (Gupta et al., in preparation(a)) which uses artificial neural networks that examine the sequence context of Asn-Xaa-Ser/Thr. NetOGlyc (Julenius et al., 2005) detects O-glycosylation of secreted and membrane-bound proteins by producing neural network predictions of mucin-type GalNAc O-glycosylation sites in mammalian proteins. O-glycans are also found on cytoplasmic and nuclear proteins and these can be detected using YinOYang (Gupta et al., in preparation(b)) which produces neural network predictions for O-ß-GlcNAc attachment sites in eukaryotic protein sequences.

3.2.3. GPI-Anchor Modification

Attachment of the glycolipid glycosylphosphatidylinositol to the C-terminus of proteins is a common post-translational modification which serves to anchor the protein to the membrane. The core glycolipid structure is conserved from protozoa to humans. There are, however, marked differences in the glycosyl side chains attached to the core glycolipid. A C-terminal propeptide, removed in the mature form, serves as the recognition sequence. The big-PI program provides predictions with separate parameters for metazoan and protozoan proteins (Eisenhaber et al., 1999) and plants (Eisenhaber et al., 2003). GPI-anchor prediction can also be performed using DGPI (Kronegg and Buloz, 1999).

3.2.4. N-Terminal Myristoylation

N-terminal myristoylation is a modification that causes the addition of a myristate to an N-terminal glycine in eukaryotic and viral proteins after removal of the initiator methionine. It is involved in directing and anchoring proteins to membranes, and as a consequence, is implicated in many essential cellular processes such as cellular regulation, signal transduction, translocation, several viral-induced pathological processes and apoptosis. The main prediction program currently used by UniProtKB curators is the NMT algorithm (Maurer-Stroh et al., 2002) which has refined the PROSITE motif for N-terminal myristoylation by analysis of known substrate sequences and kinetic data and identified three key motif regions. A more recent program, the Myristoylator (Bologna et al., 2004), makes use of neural networks to predict potential myristoylated residues.

3.2.5. Tyrosine Sulphation

Tyrosine sulphation is an important post-translational modification of proteins which go through the secretory pathway and is physiologically relevant for proteins or domains passing through or located in the Golgi lumen. The Sulfinator program (Mongatti et al., 2002) is used to detect potentially sulphated tyrosine residues. It employs four different hidden Markov models which recognise sulphated tyrosine residues located N-terminally, C-terminally, within sequence windows of more than 25 amino acids and clustered within 25 amino acid windows.

3.2.6. Phosphorylation

Protein phosphorylation at serine, threonine or tyrosine residues affects a multitude of cellular signalling processes. There are a number of methods available depending on the organism source and the enzyme catalyzing the phosphorylation. A number of PROSITE patterns have been developed to detect phosphorylation sites based on the specificity of the phosphorylating kinase. In addition, NetPhos (Blom et al., 1999) produces neural network predictions for serine, threonine and tyrosine phosphorylation sites in eukaryotic proteins.

3.3. Subcellular Location Prediction

In recent years, tools have been developed to decipher the function of a protein from its sequence, even when the most sophisticated tools for annotation transfer yield no results. Identifying the subcellular localization of a newly discovered sequence is a crucial step in narrowing down its putative function. TargetP (Emanuelsson et al., 2000) and Predotar (Small et al., 2004) are both neural network-based tools for prediction of N-terminal targeting sequences based on the predicted presence of chloroplast transit peptides, mitochondrial targeting peptides or secretory pathway signal peptides. PSORT (Nakai and Horton, 1999) uses the amino acid sequence of a protein and the type of organism from which it was obtained as the input. Based on the taxonomic origin of the protein, it searches for features characterising that taxonomic group which may help to deduce the subcellular localization of the protein using a library of known signal peptides and structural features such as topology, which may indicate whether the protein is soluble or embedded in the membrane.

4. Literature-Based Curation

The results obtained from the sequence analysis programs described above are combined with experimentally validated data extracted from scientific literature to provide a complete overview of both sequence and functional information. Results from sequence analysis programs are added only if there is information in the literature which suggests that they are valid. For example, predicted phosphorylation sites are added only if there is some evidence to suggest that a protein is either phosphorylated itself or that it belongs to a family where other members are known to be phosphorylated. Use is made of all relevant literature to ensure that the information included for each protein is complete and up-to-date, thus summarizing many pages of scientific literature into a concise yet comprehensive database record. As new information arises, entries are updated so that they continue to reflect the current state of knowledge. There are more than 1500 different journals cited in UniProt/Swiss-Prot, covering a wide range of life science publications with an average of two citations per entry and this number is higher for well-studied organisms and proteins. For example, there are an average of four publications per human entry, and UniProt entry P05067, for the human amyloid beta A4 protein which has been extensively studied, contains more than 100 references. The number of published articles is enormous and manual annotation efforts are concentrated primarily on papers reporting new functional information, structural data, post-translational modifications, polymorphisms and, where applicable, disease-causing deficiencies.

4.1. Entry Format

UniProt entries are composed of different line types, each line type having its own specified format. The large amounts of different data types found in the database are stored in a highly structured and uniform manner, which simplifies data access for users and data retrieval by computer programs. The core data, which are generally provided by the submitter of the sequence, consist of sequence data, citation information, and taxonomic data which show the biological source of the protein. Additional information added during the curation process is stored mainly in the description (DE) and gene (GN) lines, the comment (CC) lines, the feature table (FT) lines, and the keyword (KW) lines.

The description or DE line lists all the names by which a protein is known and includes standardised names assigned by official nomenclature bodies, Enzyme Commission numbers where applicable, and other synonyms from the literature. Where authoritative sources for gene names exist, such as Genew, the human gene nomenclature database (Wain et al., 2004), or the Mouse Genome Informatics group (MGI) (Eppig et al., 2005), these are used and linked. Other gene names, which have been assigned by authors, are added as synonyms. To promote interoperability, the gene identifiers assigned by genome sequencing projects are also added and are used to link to genome databases where possible.

Table 2. Comment topics used in the Swiss-Prot database

Comment topic	Description
ALLERGEN	Information relevant to allergenic proteins
ALTERNATIVE PRODUCTS	Description of the existence of protein sequences produced by alternative splicing of the same gene or by the use of alternative initiation codons.
BIOPHYSICOCHEMICAL PROPERTIES	Description of the biophysical and physicochemical properties of a protein
BIOTECHNOLOGY	Description of the biotechnological use(s) of a protein
CATALYTIC ACTIVITY	Description of the reaction(s) catalyzed by an enzyme
CAUTION	Warns about possible errors and/or grounds for confusion
COFACTOR	Description of an enzyme cofactor
DATABASE	Description of a cross-reference to a database for a specific protein
DEVELOPMENTAL STAGE	Description of the developmental-specific expression of a protein
DISEASE	Description of disease(s) associated with a deficiency of a protein
DOMAIN	Description of the domain structure of a protein
ENZYME REGULATION	Description of an enzyme regulatory mechanism
FUNCTION	Description of the function(s) of a protein
INDUCTION	Description of compound(s) which stimulate the synthesis of a protein
MASS SPECTROMETRY	Reports the exact molecular weight of a protein or part of a protein as determined by mass spectrometric methods
MISCELLANEOUS	Any comment which does not belong to any of the other defined topics
PATHWAY	Description of the metabolic pathway(s) with which a protein is associated
PHARMACEUTICAL	Description of the use of a protein as a pharmaceutical drug
POLYMORPHISM	Description of polymorphism(s)
PTM	Description of a post-translational modification
RNA EDITING	Description of any type of RNA editing that leads to one or more amino acid changes
SIMILARITY	Description of the similarity (sequence or structural) of a protein with other proteins
SUBCELLULAR LOCATION	Description of the subcellular location of the mature protein
SUBUNIT	Description of the quaternary structure of a protein
TISSUE SPECIFICITY	Description of the tissue specificity of a protein
TOXIC DOSE	Description of the lethal dose (LD) or paralytic dose (PD) of a protein

The comment or CC lines are free text comments, which are used to convey any useful information about a protein. The information in the CC lines is contained in a number of defined topics that allow the easy retrieval of specific categories of data from the database. Although free text is permissible within the comments as this is often necessary to convey detailed and complex information about a protein, a number of comments have a highly standardised syntax. A list of the currently used comment topics and their definitions is shown in Table 2.

An example of the comment lines found in a single UniProtKB/Swiss-Prot entry, P14060, is shown below:

```
CC   -!- FUNCTION: 3beta-HSD is a bifunctional enzyme, that catalyzes the
CC       oxidative conversion of delta(5)-ene-3-beta-hydroxy steroid, and
CC       the oxidative conversion of ketosteroids. The 3beta-HSD enzymatic
CC       system plays a crucial role in the biosynthesis of all classes of
CC       hormonal steroids.
CC   -!- CATALYTIC ACTIVITY: A 3-beta-hydroxy-delta(5)-steroid + NAD(+) = a
CC       3-oxo-delta(5)-steroid + NADH.
CC   -!- CATALYTIC ACTIVITY: A 3-oxo-delta(5)-steroid = a 3-oxo-delta(4)-
CC       steroid.
CC   -!- PATHWAY: Steroid biosynthesis.
CC   -!- SUBCELLULAR LOCATION: Endoplasmic reticulum and mitochondrial
CC       membrane-bound protein.
CC   -!- TISSUE SPECIFICITY: Placenta and skin. Predominantly expressed in
CC       mammary gland tissue.
CC   -!- SIMILARITY: Belongs to the 3beta-HSD family.
```

The feature table or FT lines provide position-specific data relating to the sequence. The lines have a fixed format and a defined set of feature keys which may be used. These feature keys describe domains and sites of interest within a sequence such as post-translationally modified residues, binding sites, enzyme active sites, secondary structure, and any other regions of interest. The full list of currently defined feature keys is available in the UniProtKB user manual at http://www.expasy.ch/txt/userman.txt.

The keywords are found in the keyword or KW lines of an entry. They serve as a summary of the content of an entry and assist in the retrieval of specific categories of data from the database. UniProt maintains a controlled list which currently contains approximately 860 keywords, each with a definition to clarify its biological meaning and intended usage. This list is available at http://www.expasy.org/cgi-bin/keywlist.pl and is updated on a regular basis.

4.2. Evidence Attribution

The addition of a number of qualifiers in the comment and feature table lines during the annotation process allows users to distinguish between experimentally verified data, data which has been propagated from a characterised protein based on sequence similarity, and data for which no experimental evidence currently exists (Junker et al., 1999). A recent development has been the introduction of a comprehensive evidence attribution system which allows for the linking of every piece of information in an entry to its original source. This allows users to trace the origin of all information, to differentiate easily between manual and automatic annotation, and to assess data reliability. This system has already been implemented in UniProtKB/TrEMBL and is available in the XML version of the database and through the extended view on the UniProt web site (www.uniprot.org). The system will be extended to UniProtKB/Swiss-Prot in the near future.

5. Conclusions

Literature-based manual annotation is an essential prerequisite for providing accurate and reliable data to users and for acting as a basis for the development of a range of automatic annotation methods. During the manual annotation process, experimentally validated results are combined with manually confirmed predictions from a wide range of algorithms. Due to the increasing number of prediction programs available, constant testing and comparison are essential to ensure that only those programs, which produce the most accurate results, are used.

High-quality manual annotation ensures that predictions are also of a high quality as the experimental data stored in biological databases act as a starting point for development of many novel algorithms and this, in turn, feeds back into the databases with better prediction methods available for use by curators. Manual curation ensures that experimentally valid data are combined with appropriate and high-quality prediction results to provide a complete overview of both sequence and functional information.

References

Altschul, S.F., Madden, T.L., Schäffer, A.A., Zhang, J., Zhang, Z., Miller, W., and Lipman, D.J. (1997) Gapped BLAST and PSI-BLAST: a new generation of protein database search programs. *Nucleic Acids Res.*, 25(17), 3389–3402.

Andrade, M.A., Ponting, C.P., Gibson, T.J., Bork, P. (2000) Homology-based method for identification of protein repeats using statistical significance estimates. *J. Mol. Biol.*, 298(3), 521-37.

Apweiler R. (2001) Functional information in Swiss-Prot: The basis for large-scale characterisation of protein sequences. *Brief. Bioinform.*, 2 (1), 9-18.

Argos, P., and Rao, J.K. (1986) Prediction of protein structure. *Methods Enzymol.*, 130, 185-207.

Attwood, T.K., Bradley, P., Flower, D.R., Gaulton, A., Maudling, N., Mitchell, A.L., Moulton, G., Nordle, A., Paine, K., Taylor, P., Uddin, A., and Zygouri, C. (2003) PRINTS and its automatic supplement, prePRINTS. *Nucleic Acids Res.*, 31, 400-402.

Bairoch, A., Apweiler, R., Wu, C.H., Barker, W.C., Boeckmann, B., Ferro, S., Gasteiger, E., Huang, H., Lopez, R., Magrane, M., Martin, M.J., Natale, D.A., O'Donovan, C., Redaschi, N., and Yeh, L.S. (2005) The Universal Protein Resource (UniProt). *Nucleic Acids Res.* 33, D154-D159 (2005).

Bateman, A., Coin, L., Durbin, R., Finn, R.D., Hollich, V., Griffiths-Jones, S., Khanna, A., Marshall, M., Moxon, S., Sonnhammer, E.L., Studholme, D.J., Yeats, C., and Eddy, S.R. (2004) The Pfam protein families database. *Nucleic Acids Res.*, 32, D138-141.

Benson, D.A., Karsch-Mizrachi, I., Lipman, D.J., Ostell, J., and Wheeler, D.L. GenBank. *Nucleic Acids Res.*, 33, D34-D38.

Blom, N., Gammeltoft, S., and Brunak, S. (1999) Sequence- and structure-based prediction of eukaryotic protein phosphorylation sites. *J. Mol. Biol.* 294(5), 1351-1362.

Bologna, G., Yvon, C., Duvaud, S., and Veuthey, A.L. (2004) N-terminal myristoylation predictions by ensembles of neural networks. *Proteomics*, 4(6), 1626-32.

Bru, C., Courcelle, E., Carrere, S., Beausse, Y., Dalmar, S., and Kahn, D. (2005) The ProDom database of protein domain families: more emphasis on 3D. *Nucleic Acids Res.*, 33, D212-D215.

Edgar, R.C. (2004) MUSCLE, multiple sequence aligment with high accuracy and high throughput. *Nucleic Acids Res.*, 32(5), 1792-1797.

Eppig, J.T., Bult, C.J., Kadin, J.A., Richardson, J.E., Blake, J.A., Anagnostopoulos, A., Baldarelli, R.M., Baya, M., Beal, J.S., Bello, S.M., Boddy, W.J., Bradt, D.W., Burkart, D.L., Butler, N.E., Campbell, J., Cassell, M.A., Corbani, L.E., Cousins, S.L., Dahmen, D.J., Dene, H., Diehl, A.D., Drabkin, H.J., Frazer, K.S., Frost, P., Glass, L.H., Goldsmith, C.W., Grant, P.L., Lennon-Pierce, M., Lewis, J., Lu, I., Maltais, L.J., McAndrews-Hill, M., McClellan, L., Miers, D.B., Miller, L.A., Ni, L., Ormsby, J.E., Qi, D., Reddy, T.B., Reed, D.J., Richards-Smith, B., Shaw, D.R., Sinclair, R., Smith, C.L., Szauter, P., Walker, M.B., Walton, D.O., Washburn, L.L., Witham, I.T., Zhu, Y, and Mouse Genome Database Group (2005) The Mouse Genome Database (MGD): from genes to mice - a community resource for mouse biology. *Nucleic Acids Res.*, 33, D471-D475.

Eisenberg, D., Weiss, R.M., and Terwilliger, T.C. (1982) The helical hydrophobic moment: a measure of the amphiphilicity of a helix. *Nature*, 299, 371-374.

Eisenhaber, B., Bork, P., Eisenhaber, F. (1999) Prediction of potential GPI-modification sites in proprotein sequences. *J. Mol. Biol.*, 292(3), 741-58.

Eisenhaber, B., Wildpaner, M., Schultz, C.J., Borner, G.H., Dupree, P., and Eisenhaber, F. (2003) Glycosylphosphatidylinositol lipid anchoring of plant proteins. Sensitive prediction from sequence- and genome-wide studies for Arabidopsis and rice. *Plant Physiol.*, 133(4), 1691-1701.

Emanuelsson, O., Nielsen, H., Brunak, S., and von Heijne, G. (2000) Predicting subcellular localization of proteins based on their N-terminal amino acid sequence. *J. Mol. Biol.*, 300(4), 1005-1016.

Gough, J., Karplus, K., Hughey, R., and Chothia, C. (2001) Assignment of homology to genome sequences using a library of Hidden Markov Models that represent all proteins of known structure. *J. Mol. Biol.*, 313, 903-919.

Gupta, R., Jung, E., and Brunak, S. (in preparation(a)) Prediction of N-glycosylation sites in human proteins.

Gupta, R., Hansen, J., and Brunak, S. (in preparation(b)) Identifying intracellular O-(beta)-GlcNAc `yin-yang' switches in the available human proteome.

Haft, D.H., Selengut, J.D., and White, O. (2003) The TIGRFAMs database of protein families. *Nucleic Acids Res.*, 31, 371-3737.

Harte, N., Silventoinen, V., Quevillon, E., Robinson, S., Kallio, K., Fustero, X., Patel, P., Jokinen, P., and Lopez, R. (2004) Public web-based services from the European Bioinformatics Institute. *Nucleic Acids Res.* 32, W3-W9.

Hoersch, S., Leroy, C., Brown, N.P., Andrade, M.A., and Sander C. (2000) The GeneQuiz web server: protein functional analysis through the web. *Trends Biochem. Sci.*, 25(1), 33-35.

Hulo, N., Sigrist, C.J., Le Saux, V., Langendijk-Genevaux, P.S., Bordoli, L., Gattiker, A., De Castro, E., Bucher, P., and Bairoch, A. (2004) Recent improvements to the PROSITE database. *Nucleic Acids Res.,* 32, D134-D137.

Jones, D.T., Taylor, W.R., and Thornton, J.M. (1994) A model recognition approach to the prediction of all-helical membrane protein structure and topology. *Biochemistry*, 33(10), 3038-3049.

Julenius, K., Molgaard, A., Gupta, R., and Brunak S. (2005) Prediction, conservation analysis, and structural characterization of mammalian mucin-type O-glycosylation sites. *Glycobiology*, 15(2), 153-164.

Junker, V., Apweiler, R., and Bairoch, A. (1999) Representation of functional information in the Swiss-Prot data bank. *Bioinformatics*, 15, 1066-1067.

Kanz, C., Aldebert, P., Althorpe, N., Baker, W., Baldwin, A., Bates, K., Browne, P., van den Broek, A., Castro, M., Cochrane, G., Duggan, K., Eberhardt, R., Faruque, N., Gamble, J., Diez, F.G., Harte, N., Kulikova, T., Lin, Q., Lombard, V., Lopez, R., Mancuso, R., McHale, M., Nardone, F., Silventoinen, V., Sobhany, S., Stoehr, P., Tuli, M.A., Tzouvara, K., Vaughan, R., Wu, D., Zhu, W., and Apweiler R. (2005) The EMBL Nucleotide Sequence Database. *Nucleic Acids Res.*, 33, D29-D33.

Klein, P., Kanehisa, M., and DeLisi, C. (1985) The detection and classification of membrane-spanning proteins. *Biochim. Biophys. Acta*, 815(3), 468-476.

Kronegg, J., Buloz, D. (1999) Detection/prediction of GPI cleavage site (GPI-anchor) in a protein (DGPI). *Retrieved from http://129.194.185.165/dgpi/.*

Letunic, I., Copley, R.R., Schmidt, S., Ciccarelli, F.D., Doerks, T., Schultz, J., Ponting, C.P., and Bork, P. (2004) SMART 40: towards genomic data integration. *Nucleic Acids Res.*, 32, D142-D144.

Lupas, A., Van Dyke, M., and Stock, J. (1991) Predicting coiled coils from protein sequences. *Science*, 252, 1162-1164.

Maurer-Stroh, S., Eisenhaber, B., and Eisenhaber, F. (2002) N-terminal N-myristoylation of proteins: prediction of substrate proteins from amino acid sequence. *J. Mol. Biol.*, 317(4), 541-557.

Menne K.M.L., Hermjakob H., and Apweiler R. (2000) A comparison of signal sequence prediction methods using a test set of signal peptides. *Bioinformatics*, 16(8), 741-742.

Mi, H., Lazareva-Ulitsky, B., Loo, R., Kejariwal, A., Vandergriff, J., Rabkin, S., Guo, N., Muruganujan, A., Doremieux, O., Campbell, M.J., Kitano, H., and Thomas, P.D. (2005) The PANTHER database of protein families, subfamilies, functions and pathways. *Nucleic Acids Res.*, 33, D284-D288.

Moeller, S., Croning, M.D.R., and Apweiler, R. (2001) Evaluation of methods for the prediction of membrane spanning regions. *Bioinformatics* 17(7), 646-653 (2001).

Mongatti, F., Gasteiger, E., Bairoch, A., and Jung, E. (2002) The Sulfinator: predicting tyrosine sulfation sites in protein sequences. *Bioinformatics*, 18(5), 769-770.

Mulder, N.J., Apweiler, R., Attwood, T.K., Bairoch, A., Bateman, A., Binns, D., Bradley, P., Bork, P., Bucher, P., Cerutti, L., Copley, R., Courcelle, E., Das, U., Durbin, R., Fleischmann, W., Gough, J., Haft, D., Harte, N., Hulo, N., Kahn, D., Kanapin, A., Krestyaninova, M., Lonsdal,e D., Lopez, R., Letunic, I., Madera, M., Maslen, J., McDowall, J., Mitchell, A., Nikolskaya, A.N., Orchard, S., Pagni, M., Ponting, C.P., Quevillon, E., Selengut, J., Sigrist, C.J., Silventoinen, V., Studholme, D.J., Vaughan, R., and Wu, C.H. (2005) InterPro, progress and status in 2005. *Nucleic Acids Res.* 33, D201-D205.

Nakai, K., Horton, P. (1999) PSORT: a program for detecting sorting signals in proteins and predicting their subcellular localization. *Trends Biochem. Sci.*, 24(1), 34-36.

Neilsen, H., Engelbrecht, J., Brunak, S., and von Heijne, G. (1997) Identification of prokaryotic and eukaryotic signal peptides and prediction of their cleavage sites. *Protein Engineering*, 10, 1-6.

Notredame, C., Higgins, D., and Heringa, J. (2000) T_Coffee: a novel method for multiple sequence aligments. *J. Mol. Biol.*, 302(1), 205-217.

Notredame, C. (2002). Recent progress in multiple sequence alignment: a survey. *Pharmacogenomics* 3(1), 131-44.

Pearl, F., Todd, A., Sillitoe, I., Dibley, M., Redfern, O., Lewis, T., Bennett, C., Marsden, R., Grant, A., Lee, D., Akpor, A., Maibaum, M., Harrison, A., Dallman, T., Reeves, G., Diboun, I., Addou, S., Lise, S., Johnston, C., Sillero, A., Thornton, J., and Orengo, C. (2005) The CATH Domain Structure Database and related resources Gene3D and DHS provide comprehensive domain family information for genome analysis. *Nucleic Acids Res.*, 33, D247-D251.

Pearson, W.R. and Lipman, D.J. (1988) Improved tools for biological sequence comparison. *Proc. Natl. Acad. Sci. U.S.A.*, 85(8), 2444-8.

Riley, M.L., Schmidt, T., Wagner, C., Mewes, H.W., and Frishman, D. (2005) The PEDANT genome database in 2005. *Nucleic Acids Res.*, 33, D308-D310.

Small I., Peeters N., Legeai F., and Lurin C. (2004) Predotar: a tool for rapidly screening proteomes for N-terminal targeting sequences. *Proteomics*, 4(6), 158115-90.

Sonnhammer, E.L., Heijne, G., and Krogh, A. (1998) A hidden Markov model for predicting transmembrane helices in protein sequences. *Proc. Int. Conf. Intell. Syst. Mol. Biol.*, 6, 176-182.

Sturrock, S.S., and Collins, J.F. (1993) *MPsrch V1.3 User Guide*. Biocomputing Research Unit, University of Edinburgh, Edinburgh.

Tateno, Y., Saitou, N., Okubo, K., Sugawara, H., and Gojobori, T. (2005) DDBJ in collaboration with mass-sequencing teams on annotation. *Nucleic Acids Res.*, 33, D25-D28.

Thompson, J.D., Higgins, D.G., and Gibson, T.J. (1994) CLUSTAL W: improving the sensitivity of progressive multiple sequence alignment through sequence weighting, position-specific gap penalties and weight matrix choice. *Nucleic Acids Res.*, 22(22), 4673-80.

Wain, H.M., Lush, M.J., Ducluzeau, F., Khodiyar, V.K., and Povey, S. (2004) Genew: the Human Gene Nomenclature Database, 2004 updates. *Nucleic Acids Res.*, 32, D255-D257.

Wu, C.H., Nikolskaya, A., Huang, H., Yeh, L.S., Natale, D.A., Vinayaka, C.R., Hu, Z.Z., Mazumder, R., Kumar, S., Kourtesis, P., Ledley, R.S., Suzek, B.E., Arminski, L., Chen, Y., Zhang, J., Cardenas, J.L., Chung, S., Castro-Alvear, J., Dinkov, G., and Barker, W.C. (2004) PIRSF: family classification system at the Protein Information Resource. *Nucleic Acids Res.*, 32, D112-D114.

In: In Silico Genomics and Proteomics
Editors: N. Mulder and R. Apweiler, pp. 71-78

ISBN 1-59454-995-8
© 2006 Nova Science Publishers, Inc.

Chapter 6

Automated Annotation of Proteins

Ernst Kretschmann and Daniela Wieser

European Bioinformatics Institute, Wellcome Trust Genome Campus, Hinxton,
Cambridge, UK, CB10 1SD

Abstract

Automated annotation refers to the process in which annotation is added to protein
database entries without the assistance of human experts. This chapter makes clear why
automated annotation is beneficial and it explains the most important aspects of the
automated annotation methods developed and applied at the European Bioinformatics
Institute (EBI). The prediction systems RuleBase and Spearmint are explained and the
contradiction system Xanthippe is introduced. The latter is used to filter false positive
predictions produced by the prediction systems or to detect erroneous imports from other
databases. All systems are applied at the EBI on a regular basis in order to predict
annotations for the proteins in the TrEMBL section of the UniProt Knowledgebase.
Therefore, a highly efficient automated annotation pipeline was developed, which will be
introduced at the end of this chapter.

1. Introduction

In recent years, high-throughput genome sequencing projects have provided the scientific
world with a wealth of decoded protein sequences. Cataloguing the function of these proteins
requires a literature curation process in which all available publications about a protein are
manually checked to associate it with a particular annotation. As a consequence of this time-
consuming process the proportion of well-characterized proteins compared to the amount of
decoded sequences is decreasing. One way to keep up with the challenge of characterizing
the huge amount of protein sequencing data is to develop and to use automated annotation.

Automated annotation refers to the process in which annotation is added to protein
database entries without the assistance of human experts. We use automated annotation for

the UniProtKB/TrEMBL (UniProt Knowledgebase) section [1] to generate annotation items in the often sparsely annotated protein entries. To do this in a highly efficient way we have developed mechanisms for the three basic steps into which the automated annotation procedure falls. Data-preparation arranges all needed data items in a data warehouse structure to allow efficient access to the data sources. In order to create annotation rules we use parallelized learning methods to extract knowledge about core data dependencies in the prepared protein data. This knowledge is applied in a high-throughput data generating application pipeline to currently about 2 million protein sequences. This chapter will give an overview of automated annotation applied to the UniProtKB [1] and it will summarise the main steps involved in the automated annotation pipeline. We will give a brief history of automated annotation attempts, describe the current state of the art, and give an outlook on what can be achieved in the near future.

2. Why Automated Annotation Is Needed

An abundance of protein data of diverse usefulness is available in various protein data resources. One of the most comprehensive sources is the curated UniProtKB/Swiss-Prot data set [1], which gives an overview of many relevant properties of a protein in a standardised way. Each protein runs through a sequence of annotation steps performed by a team of human experts. This process delivers high data quality but does not scale to the amount of incoming data. Years ago the objective to provide this high quality standard to all known proteins had to be abandoned and TrEMBL was created as a resource for all proteins, which had not been manually processed. At this time of writing, the ratio of well-annotated proteins against uncharacterised ones is about one to ten. If you happen to read this text in a couple of years, you might look back in amusement: This gap doubles every 18 months and we will reach 1:100 probably around 2008. Similarity searches against UniProt will result in an abundance of hits to the UniProtKB/TrEMBL section but almost none to the UniProtKB/Swiss-Prot section. The similarity search can be done against Swiss-Prot only, but Swiss-Prot unfortunately is not a representative collection of protein data and the closest relatives to the sequence of interest might be missed. Users will have to examine the results of their similarity searches, look at the taxonomic range of organisms in their result, browse through the family classifications and interpret all this to come to a conclusion about the possible properties of these proteins.

This interpretation is what automated annotation is supposed to do on the fly and without human interaction. Only if the data in TrEMBL gives some satisfying automatically generated annotation, which might not be comprehensive but reliable to a certain degree, will similarity searches continue to make sense.

3. From Core Data to Annotation

Some of the data items in Swiss-Prot are hard, fundamental and non-negotiable facts. Proteins are identified in certain organisms and have sequences that were determined by scientific experiments. Given the sequence, there are hits to sequence patterns calculated such as profiles from PROSITE [5] or HMMs from Pfam [2], and given these hits the protein is classified into families or domains in InterPro [7]. Most of this data is automatically added and available for any protein sequence in UniProt - but we do not consider this annotation. Annotation is the added value that is usually produced by the literature curation process. These are all the soft, interpretative and descriptive items such as the assigned protein name that is found in the description line (DE) of Swiss-Prot entries. Additional information can be found in the Keywords (KW), Comments (CC) and Feature (FT) parts of the protein entry. Automated annotation is about the question if and how annotation data can be produced automatically without the involvement of an expert.

Fortunately, there is a link between core data and annotation. The following examples will briefly illustrate this dependency:

- All the 68 proteins belonging to the InterPro family of the Connexins (InterPro IPR000500) are annotated with the keywords *"Gap junction"* and *"Transmembrane"* this is a good indication that all the proteins of this family, no matter if they reside in Swiss-Prot or TrEMBL, should be annotated with them.

- Most of the eukaryotic Swiss-Prot proteins from InterPro IPR000685 (434 out of the 436) have the *"Chloroplast"* keyword annotation - all the 41 non-eukaryotic ones do not have it. This makes a good annotation rule: If it is in InterPro IPR000685 and it is eukaryotic, it should have the keyword *"Chloroplast"*. Note, that for the last example no biological argument has been used, it is based purely on the data distribution in Swiss-Prot.

- All 41 members of the *"Kringle"* family IPR000001, which have a hit to PROSITE PS00135 have the keyword *"Serine protease"* annotated. All the 18 others do not, which is itself a good annotation rule.

Core data apparently indicates annotation data, so how can these data dependencies be detected? There is a manual approach to do this, which is actually employed in the construction of the RuleBase [3] annotation rule set. The InterPro families are inspected manually, data dependencies are extracted, and the result is translated into annotation rules. The advantage of this approach is obvious: The protein family can be checked, in the case of the example we were dealing with the group of the *"Ribulose bisphosphate carboxylase, large chain"* proteins, for which we find that *"in plants, the large subunit is coded by the chloroplastic genome"*. One of the two examples from Swiss-Prot, where the *"Chloroplast"* annotation is missing is from *"Cyanophora paradoxa"*, an organism that has Cyanelle organelles instead of chloroplasts. The other is annotated as being *"very hypothetical"*, which is an indication that very little about this protein is known and annotated - provided it actually exists. Taken everything into account, there is a good biological argument for the rule to be exported.

However, there are problems with this approach. First, it is time consuming to check for the biological reasons behind an annotation rule. Second, the data dependencies are not always obvious. Third, there is an overwhelming abundance of protein families, which is simply too much data to be mined comprehensively in a purely manual way. As for the annotation of protein entries, a manual-only approach has its limits because it fails to scale to the amount of incoming sequence data.

There are automatic alternatives, though. Standard data mining algorithms can be used to detect the data dependencies and produce annotation rules automatically. The most powerful in terms of speed, reliability and readability of the results is the C4.5 algorithm (and its commercial successors), which takes a data dependency matrix as input and generates a decision tree out of it. The exact mechanism of how this works goes beyond the scope of this book but can be found in Quinlan's "C4.5: Programs for Machine Learning" [8]. A short illustration about how the automated annotation systems can profit from data mining algorithms is given below.

The Spearmint [6] system uses the C4.5 decision tree algorithm to automatically classify the proteins of the given training set into positive and negative examples depending on the presence of a particular annotation. An example for a generated decision tree is given in Fig. 1. It consists of a root node, several condition nodes and leaf nodes. The root and condition nodes represent sequence patterns that can be computed for all sequences by secondary databases like Pfam or ProDom [9], whereas the leaf nodes hold the annotation. The Spearmint system extracts rules from those parts of the tree that describe the positive instances of the training set. About 10.000 rules are currently generated this way on a fortnightly basis and are used to classify uncharacterized sequences.

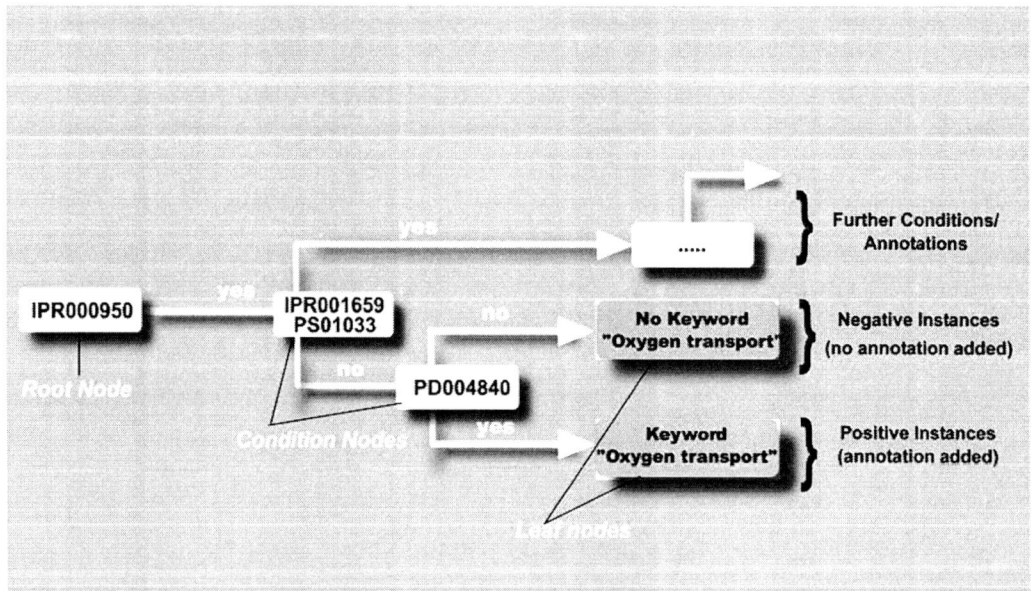

Fig. 1. Decision tree that classifies proteins belonging to IPR000950. The Spearmint prediction rule is: "Add the Keyword "Oxygen transport", if the protein belongs to IPR000950 (InterPro), but neither to IPR001659 nor to PS01033 (PROSITE) and if it has a hit to PD004840 (ProDom)"

The algorithm can be used to export predictions for all possible annotations found anywhere in Swiss-Prot. Sometimes these are supported by convincing statistics, so that manual biological checks can be avoided. In other cases the statistics are insufficient: Many instances in UniProt/Swiss-Prot do not conform to the rule or there are too few examples to support it statistically. A decision has to be made, when a rule is good enough to be exported and applied without the necessity of repeatedly checking the annotation by human experts. This is done with the help of a statistical equation that calculates the number of positive and negative examples in the training set and produces a normalized value, with which good rules can be separated from bad ones. The advantage of this selection procedure is that the produced annotation quality can be configured. By selecting only the rules supported by unambiguous statistics, the predictions can be tuned to be highly reliable. By also exporting less convincing rules, a greater recall rate can be achieved for the price of producing some erroneous annotation.

4. Preventing Erroneous Annotation

Erroneous annotation is produced by any kind of annotation mechanism including manual and automatic approaches and there is unfortunately no way to avoid this completely. Some uncharacterized proteins are very similar to known examples but still behave differently in their context. There are cases where the signature hits of the protein sequence clearly suggest properties that these proteins lack. Transferring annotation from the well-investigated samples to those that do not stick to the framed annotation rules will produce "over predictions" or outright errors. This applies not only for exceptions of an annotation rule, but also for proteins that are mistakenly associated with signature hits (false positive hits). Additional errors are introduced when proteins have false negative signature hits, i.e. one ore more patterns should hit the protein sequence, but they do not.

Many of the above mentioned errors can be detected by using biological reasoning and some even by statistical approaches using data mining techniques. The idea for all approaches is to search for correlations between core data and annotation that could be produced in the automated annotation pipeline but were never observed in the well-characterized training set. This technique establishes annotation exclusion rules for these combinations, which can be applied as quality assurance subsequent to the prediction steps. A warning is given, or in some cases an annotation is removed, whenever it was predicted but also fulfils one or more exclusion rule. The exclusions are simple and obvious in many cases. For instance, bacteria lack a cell nucleolus and hence bacteria do not express nuclear proteins. Yet, 12 bacterial examples in TrEMBL are annotated with the *"Nuclear protein"* keyword and the simple exclusion rule: "If a protein is from Bacteria, disallow the annotation of the keyword *"Nuclear protein"*, improves these entries.

Even though the simple mapping approach is capable of detecting the most serious annotation errors, the approach is problematic. Since most of the possible annotation - core data combinations do not occur in the Swiss-Prot database, billions of exclusion rules can be exported and the application of the rules to currently 2 million protein entries in a reasonable time frame is impossible. The set of possible exclusion rules can be reduced to a smaller set, which only contains exclusions likely to actually occur. These prevent the most frequently made errors. One example of this idea is the text editor, with which this chapter is written: whenever the combination hte is typed, it automatically converts it to the, which is a frequent error but it deos nto do it fro lal tyops, only for the most likely ones. For protein annotation, likely mistakes are those, which are derived from proteins having similar core properties to, for instance, the ones in the same protein family. Exclusion rules affecting these annotations are likely to spot more errors than those contradicting random combinations. From the above example, it can be concluded that belonging to InterPro IPR000001 is not sufficient to be able to add the *"Serine protease"* keyword. There also needs to be the PROSITE PS00135 hit. Indeed, there are eight entries in TrEMBL, for which there is a *"Serine protease"* keyword although the PROSITE PS00135 hit is missing. Chances are high that this annotation is erroneous. The exclusion rules spotting the frequently annotated errors can be produced by a similar approach to the one used to export Spearmint annotation rules. Wieser *et al.* [10] describe this process in detail and call their exclusion rule set Xanthippe. They show, that about 50% of all annotation errors produced by Spearmint can be avoided using Xanthippe. Like all data mining based routines, this quality improvement comes at a price: some actually correct annotations are contradicted and a recall - precision compromise has to be made. As the Xanthippe rule-set is created fully automatically and configurable, the actual detection rate can be chosen. The higher this rate, the more actually correct annotations will be contradicted. Fig. 2 illustrates the current performance (beginning of 2005) of all three automated annotation systems, i.e. Spearmint, RuleBase and Xanthippe. A higher total number of keyword predictions would lead to an increased number of false positive predictions, but also to a higher error detection rate.

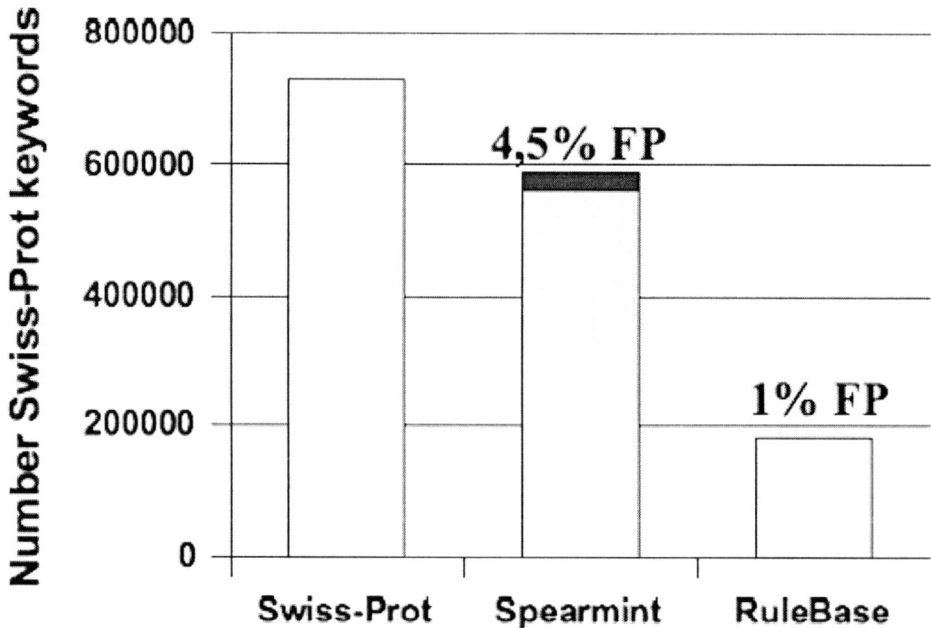

Fig. 2. Quality measurements for Spearmint and RuleBase. The diagram shows true positive (light grey) and false positive keyword predictions (dark grey) of the systems once applied to Swiss-Prot. The contradiction system Xanthippe detects 30% of the Spearmint false positive predictions and 10% of the RuleBase false positive predictions (not shown in the diagram)

5. The Automated Annotation Pipeline

In this chapter we leave the shiny data-mining topic and delve into the mundane tasks required to export data mining results in reasonable time. The very best data mining results are of very limited value, unless they can be applied to generate annotation in uncharacterised protein entries. There is a new UniProt release every two weeks, so rules have to be generated and applied inside this very limited time frame. One very time-consuming step in this process is the generation of the annotation rules. Fortunately this step is detached from the production pipeline and can be done at any time. Yet, it is time consuming to prepare the data for the mining algorithms: For all InterPro groups all UniProt/Swiss-Prot have to be extracted to serve as training sets. For this purpose InterPro and UniProt data have to be assembled in one database instance that guarantees read optimized access, in short, a data warehouse. A data warehouse component is a crucial part of any data mining routine because it facilitates the operation and the extension of all data mining routines. There are different approaches to building such a warehouse, for the automated annotation pipeline a component was chosen that is not updateable and is refilled biweekly. This way the data content is representing a snapshot in time and is in itself consistent. There is no such thing as an update, such as the addition of proteins to an InterPro group, during the data mining procedure that could possibly lead to inconsistencies in the produced annotation rules.

Ironically, the actual step of mining the data is the least time consuming. Filling the warehouse with content and preparing the data for the mining routine takes much longer, but the most time critical step is the production of automated data for uncharacterised proteins. There are about 100.000 rules exported from 12.000 InterPro groups, which have to be applied to nearly 2.000.000 protein sequences. If the application of one rule against one protein entry took as little as 1ms - which is unrealistically optimistic - it would take about 6 years to perform an automated annotation run. Fortunately, the annotations of individual proteins do not depend on each other and hence the annotation process can be performed in parallel threads. Still, this approach alone would result in massive hardware costs to reduce runtimes to sensible time scales. The central optimization facility that helps condense this application into a couple of minutes is the reduction of the complexity class from N^2 to $Nlog(N)$. For this, the annotation rules are separated into classes: one class is the set of rules that require a protein to be in InterPro IPR000001 to possibly apply, the second class requires InterPro IPR000002 etc. For each protein, the membership to InterPro domains and families is analysed before rule application and only those rules that actually stand a chance to hit are executed. Since most proteins are only hit by a very small number of annotation rules, this procedure in combination with a parallel approach facilitates the application.

6. Conclusion

In summary, automated annotation is only partly about sophisticated algorithms. Granted, the better these are the better the results become, but there is also a lot to invest in questions on how to store and prepare training data. Dealing with massive data sets often requires optimizations and making compromises between the added value produced and calculation times required. Only a balanced investment of development time into all aspects of automated annotation facilitates an iterative improvement of the product. Hence, in the near future we will provide predictions about GO-Terms [4], which will add thousands of new annotation rules. To do this, GO annotation will have to be included in the warehouse, prediction algorithms will be evaluated in cross-validations and the application pipeline will be profiled to be able to present these predictions to our users.

References

[1]   Apweiler R., Bairoch A., Wu C.H., Barker W.C., Boeckmann B., Ferro S., Gasteiger E., Huang H., Lopez R., Magrane M., Martin M.J., Natale D.A., O'Donovan C., Redaschi N., Yeh L.L. UniProt: the Universal Protein knowledgebase. *Nucleic Acids Res.*,2004, 32:D115-D119.

[2]   Bateman A., Coin L., Durbin R., Finn R. D. , Hollich V., Griffiths-Jones S., Khanna A., Marshall M., Moxon S., Sonnhammer E. L. L., Studholme D. J., Yeats C. and Eddy S. R. The Pfam Protein Families Database. *Nucleic Acids Research*, 2004, 32, D138-D141.

[3]   Biswas M., O'Rourke J.F., Camon E., Fraser G., Kanapin A., Karavidopoulou Y., Kersey P., Kriventseva E., Mittard V., Mulder N., Phan I., Servant F., Apweiler R. Applications of InterPro in protein annotation and genome analysis. *Briefings in Bioinformatics 3(3):*,2002, 285-295.

[4]   Camon E., Barrell D., Lee V., Dimmer E., Apweiler R. The Gene Ontology Annotation (GOA) Database - An integrated resource of GO annotations to the UniProt Knowledgebase, 2004, *In Silico Biology 4(1): 5-6.*

[5]   Falquet L., Pagni M., Bucher P., Hulo N., Sigrist CJ., Hofmann K., Bairoch A. The PROSITE database, its status in 2002. *Nucleic Acids Res.* 2002 Jan 1;30(1):235-8.

[6]   Kretschmann E., Fleischmann W., Apweiler R. Automatic rule generation for protein annotation with the C4.5 data mining algorithm applied on SWISS-PROT. *Bioinformatics*, 2001, 17:920-926.

[7]   Mulder,N.J. et al. InterPro, progress and status in 2005. *Nucleic Acids Res.* 2005; 33:D201-5.

[8]   Quinlan. R. C4.5: Programs for Machine Learning, Morgan Kaufmann, 1993.

[9]   Servant F, Bru C, Carrïre S, Courcelle E, Gouzy J, Peyruc D, Kahn D. ProDom: Automated clustering of homologous domains. *Briefings in Bioinformatics*. Vol 3, 2002, no 3:246-251.

[10]   Wieser D., Kretschmann E., Apweiler R. Filtering erroneous protein annotation. *Bioinformatics*, 2004, 20:i342-i347.

In: In Silico Genomics and Proteomics
Editors: N. Mulder and R. Apweiler, pp. 79-88

ISBN 1-59454-995-8
© 2006 Nova Science Publishers, Inc.

Chapter 7

Guilt by Association: Using Protein Interactions and Expression Patterns to Predict Protein Function

Sandra Orchard* and Henning Hermjakob

EMBL Outstation – European Bioinformatics Institute, Wellcome Trust
Genome Campus, Hinxton, Cambridge, UK

Abstract

It is possible to annotate novel proteins by homology to closely related homologues and orthologues. These may be recognised by sequence similarity or by using patterns, motifs or Hidden Markov models directed at recognising specific regions of the protein sequence. For proteins with no notable homology to known sequences, other techniques must be used to predict cellular function and the processes in which they are involved. An understanding of the interactions made by these molecules, when combined with well-annotated protein expression data, can give valuable clues to assist in elucidating molecular function.

Introduction

The explosive increase in genome sequencing has resulted in the identification of an ever-increasing number of proteins for which the sequences have been determined or predicted. Tools such as InterPro [1] use regular expressions, motifs and Hidden Markov models, built from seed alignments of proteins of known function, to predict the role close homologues play within a cell. Similarly, sequence comparison to well annotated proteins in UniProtKB [2] may also give clues to protein function. However, there still remain many

---

*To whom correspondence should be addressed.Tel: +44 (0)1223 494 675; Fax: +44 (0)1223 494 468; Email: orchard@ebi.ac.uk

proteins with no close analogues and which are not members of any known protein families. For these proteins, other techniques need to be used to give an indication of the function of these molecules. The interactions an individual protein makes with other proteins, with nucleic acids, lipids or small molecules can all give information as to the physiological role of the molecule within a cell. Protein interaction data has been reported in the literature for many years and is now being generated in increasingly large volumes. Such data is now being collected in molecular interaction databases which are publicly available for searching and data retrieval. However, in many cases the techniques that are used to generate such data remove the protein from the natural environment of the cell and from the constraints of subcellular location or the effects of the cell cycle or of external stimulants on protein expression patterns. It is therefore essential that such data be seen in conjunction with protein expression data such as is generated by proteomics experiments. Again, there is much of this data available in the literature but, until very recently, there have been no central repositories in which it can be stored, accessed and cross-referenced to protein sequence databases or databases of molecular interaction. The development of such repositories is now ongoing and new possibilities are opening up in the field of protein annotation.

Intact: A Molecular Interaction Database

The IntAct molecular interaction (www.ebi.ac.uk/intact) database has been developed for the deposition and storage of protein:protein, protein:nucleic acid and protein:small molecule information [3]. IntAct also supports a team dedicated to the curation of experimental data already available in the literature, this effort is currently strongly biased towards the collection of protein:protein interaction data. The IntAct data model has three main components: Experiment, Interaction and Interactor. An Experiment groups a number of Interactions, usually from one publication, and classifies the experimental conditions in which these Interactions have been generated. An Experiment may have only a single interaction, or hundreds of interactions in the case of large-scale experiments. An Interactor is a biological entity participating in an Interaction, usually either a protein, a nucleic acid sequence, or a small molecule. An Interaction contains one or more Interactors participating in the Interaction. The representation of interactions is not limited to binary interactions; data on multi-protein interactions, e.g. the results of tandem affinity purification experiments [4] can be represented as one interaction, without artificially splitting them up into several binary interactions.

Molecule identifications, molecular binding features and experimental techniques all need to be described in an unambiguous manner such that both the laboratory scientist and a computer program can search and access the data, understand the results and be able to compare this with locally derived data or with data downloaded from other sources. IntAct achieves this by using extensive cross-referencing to key source databases and by the use of controlled vocabularies to describe any area of the database available to search.

Protein identification is dealt with by using the stable protein accession numbers generated by the UniProtKB database. In the rare cases where a protein is described in the absence of sequence information, for example a mouse antibody is used in a rat cell line but the rat protein has yet to be sequenced, a protein entry can be generated within IntAct until such time as the sequence becomes publicly available in UniProtKB. Similarly nucleic acid identifiers will be annotated by DDBJ/EMBL/Genbank identifiers [5] and small molecules using ChEBI descriptors and accession numbers [6]. Interacting domains are identified by cross-referencing to InterPro with residue numbering directly linked to that of the given sequence used by the corresponding UniProtKB entry, residue renumbering is undertaken in line with sequence updates in the UniProtKB database.

All experimental detail and additional molecule feature information is described, wherever possible, by the use of controlled vocabularies. The use of ontologies and vocabularies are proving essential as biological data and the corresponding terms to describe such data proliferate. Single biological process can now be described by many different terms and their synonyms which compromises the ability of search engines to identify complete data sets. For example, to the human eye the phrases yeast-2-hybrid, yeast 2 hybrid and Y2H all describe the same process. To a text-mining program they are very different and the program would require special instruction to group such data into a single set. By using a single term throughout the database to describe this experimental technique, and then making that term known to the user, data can be searched, sorted and filtered without suffering data loss due to poor use of terminology. Where possible, existing reference systems are used, such as the NCBI taxonomy database [7] to delineate taxonomy or the Gene Ontology (GO) [8] to describe complex function, involvement in biological processes and subcellular location.

For a number of attributes specific to molecular interactions, new controlled vocabularies have been developed in the IntAct project, together with extensive definitions and cross-references. The vocabularies are used to describe many aspects of the experimental procedures and interaction details, for example "participant type", "interaction type", "interaction detection method", "participant detection method", "feature type" and "feature detection method". In specific cases, these vocabularies are closely linked to, and cross reference, other specialist databases. For example, the post-translational modifications described in "feature type" are all cross-referenced to the RESID database [9] where such modifications are described in fuller detail. All these controlled vocabularies have been made publicly available.

Annotation Using Protein Interaction Data

If two proteins interact in a physiologically relevant cellular environment, then they must share the same subcellular location during a common period of time within the cell cycle or differentiation stage. It is also reasonable to expect that they are involved in the same biological process, although an interaction with a ribosome or proteosome protein may only indicate the biosynthesis or degradation of the protein of interest. By examining the interactions a protein has been shown to make, valuable information may be gained as to its function. In the example shown in Fig. 1, the protein of interest, Q8NI08, has yet to be annotated within UniProtKB/Swiss-Prot at the time of writing and is only known by the name used by the original sequence submitter, ERAP140. The graphical view within IntAct displays all GO terms which are annotated to proteins in the displayed interaction network. Any of these GO terms can be selected, and all proteins which have this GO term or any of its child terms annotated, will be highlighted. This functionality provides a quick method to interactively explore the functional context of proteins. ERAP140 interacts with a number of proteins, which all share the GO annotation for the subcellular location "Nucleus" suggesting ERAP140 may also be a nuclear protein. Many of these proteins are also annotated as transcription factors or with roles in transcriptional control, again suggesting that ERAP140 may also be involved in the process of transcription. A look at the reference that accompanied the original sequence submission to the nucleotide database confirms this protein to be a nuclear receptor coactivator that mediates oestrogen-receptor induced gene expression.

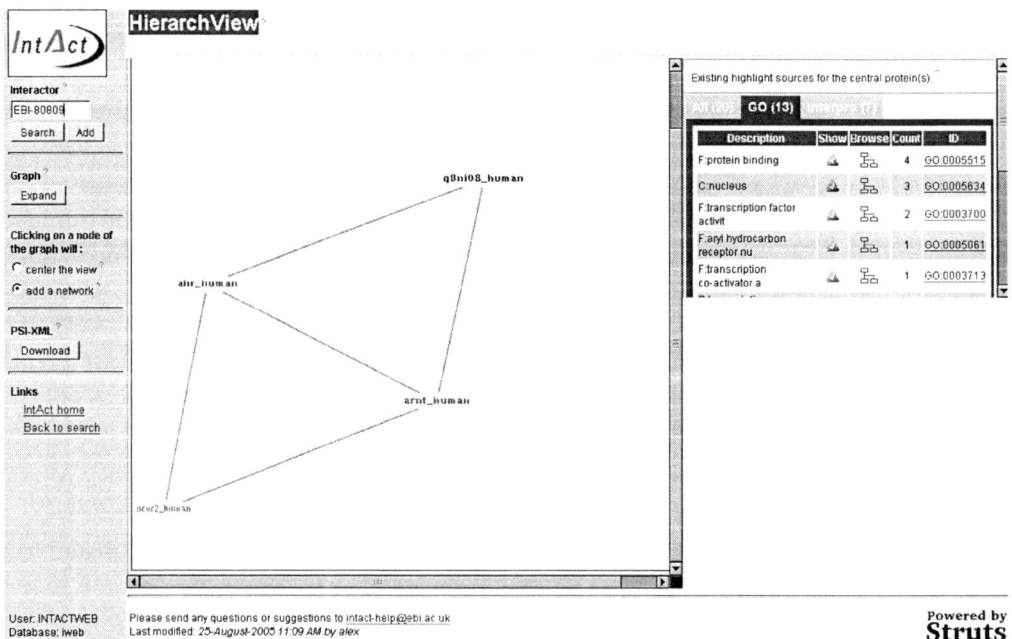

Fig 1. HierachView graphical display showing the interacting partners of Q8NI08

This process may even be used as a means of automatically adding information to larger sets of proteins, or for improving existing annotation, by looking at common annotation which is shared by groups of interacting proteins. For example, a group of stably interacting proteins may all share the common GO component term GO:0005835 (fatty-acid synthase complex). A machine-learning process could then annotate all these proteins to the GO process GO:0004312 (fatty acid synthase activity), which may be present on some but not all the proteins in the complex. A comparison across UniProt may lead to the addition of the common annotation block "Fatty acid synthetase catalyzes the formation of long-chain fatty acids from acetyl CoA, malonyl-CoA and NADPH" and the addition of the EC number EC 2.3.1.86 to all the proteins in this complex. Proteins which have been annotated to a higher level term, for example GO:0016740 (transferase activity), may have this less granular annotation updated.

The Limitations of Protein Interaction Data

Protein interaction data is produced by many different experimental techniques, all of which are capable of answering many but not all the questions asked of it. Coimmunoprecipitation, for example, can pull down complexes formed in a particular cell or organelle under physiological conditions but will bring down all the proteins and protein complexes capable of interacting with a particular bait. The technique does not separate binary from indirect interactions. Conversely, yeast 2 hybrid has been widely used because it is particularly appropriate for library screening for the initial identification of binary interacting partners and can easily be run in high throughput mode. However, it takes the proteins away from the natural cell environment and will demonstrate interactions between proteins, which under physiological conditions, would either be temporally or spatially separated and the interaction would never occur.

For this reason, it would be sensible to look at interaction data in conjunction with protein expression data, which has been well annotated to include available detail of cell type, environment and stage of the cell cycle as well as full detail of where in the cell the protein has been expressed. The advent of high throughput proteomics has allowed the identification of ever increasing numbers of proteins, often isolated from very specific cellular organelles, during a particular phase of a cell cycle or following cell activation with a particular agonist. However, the publication of these lists of protein identifications has proved problematic, with long lists often referred to but stored in supplementary materials, on web sites or on author-maintained databases. However, repositories are now coming on-line in which this data can be stored and subsequently accessed.

PRIDE: A Proteomics Identifications Database

The PRIDE PRoteomics IDEntifications database (www.ebi.ac.uk/pride) is a centralized, standards compliant, public data repository for proteomics data [10]. It has been developed to provide the proteomics community with a public repository for protein and peptide identifications together with the evidence supporting these identifications. The PRIDE project consists of a number of distinct parts. The PRIDE core libraries contain an object model of the PRIDE data structure and allow the programmer to interact seamlessly and effortlessly with the PRIDE XML format and relational database implementation. The PRIDE web libraries provide a web-based view of an underlying reference database and use the PRIDE core libraries for data access. Query results from the web can be sent in PRIDE XML format or in HTML after XML Stylesheet Language (XSL) transformation of the XML.

In PRIDE, one or more experiments are contained in the root tag 'ExperimentCollection'. The top-level structure of an experiment, consists of seven conceptually distinct parts. The first of these is the experiment accession number, assigned after successful submission of an experiment. The experiment accession number would be the data element of choice for inclusion in papers as the PRIDE reference because of its conciseness and since interested readers can easily use it to quickly retrieve all relevant data from the PRIDE web interface. The second part contains meta-data about the experiment: a descriptive title, contact person and/or address, a short label, a description and finally location information. The third element concerns the sample studied. It consists of a description field and an attribute list. Protocol information constitutes the fourth part of an experiment. As well as a description and attribute list, it also holds one or more sections about the mass spectrometer(s) used, including manufacturer, model, source and analyzer information which can be further supplemented through an attribute list. The fifth part details the information derived from the mass spectrometer. This section holds the MS coefficient (e.g. MS^2, MS^3), peak lists, optional raw data references, comments and an attribute list. The most intricate subsection of an experiment is the sixth part and deals with the identifications obtained from the data specified in part five. Identifications have been split in two different types: two-dimensional polyacrylamide gel electrophoresis (2DPAGE) based identifications and non-gel based identifications. The shared elements are wrapped up in an abstract ancestor element called 'IdentificationType'. The additional information for 2D-PAGE based identifications centres on protein-related data gathered during the gel separation phase, whereas the gel-free identifications typically require more information about the effective identification score and threshold (if available). Finally, the seventh part is not restricted to the experiment level but can be found in many of the smaller branches as well. This is the 'AttributeList' which represents a list of attributes, to be keyed from controlled vocabularies, allowing an extremely flexible way of integrating additional information into the core schema without sacrificing the structure of the whole. In fact, the PRIDE schema presents a minimum minimorum of information about protein identifications in present day proteomics.

PRIDE may now also be utilized as an annotation tool, since experimental detail added to each entry through the use of controlled vocabularies such as GO, MESH or custom-written if necessary, allows the submitters and curators to add information such as the exact clinical conditions under which the tissue was isolated or specific subcellular locations that the proteins were extracted from. The researcher may then ask questions such as "show me in which tissue my protein(s) have previously been observed" or "in which context has a similar protein set to my reference set been observed". In the latter case, proteins could be grouped by function or process by using GO terms, by stage in a cell-cycle or involvement in a particular pathway. Common annotation blocks for proteins grouped by expression may then be automatically extracted from the underlying UniProtKB entries and the extensive cross-referencing of each UniProtKB entry also utilized, for example to the Reactome pathways database [11].

Data Standardisation

The establishment of public domain repositories in which protein interaction and expression can be stored is a major step forward in the study of protein science but isolated, stand-alone databases are, in themselves of limited use. The community benefits greatly if such databases are capable of interchanging data such that a common pool of high quality information is available from a single source. When data is scattered across a number of separate sources, and each maintains its own individual data format, the user is forced into time-consuming and difficult parsing exercises before they can generate a single non-redundant dataset.

In April 2002, the Protein Standards Initiative (PSI) committee was formed by the Human Proteome Organisation (HUPO) [12] and tasked with standardising data formats within the field of proteomics. Public domain databases could be established or adapted for data deposition using these formats and data could then be exchanged between and downloaded from existing databases using the formats as a common exchange mechanism. In the interests of reducing this task to one of manageable proportions, the PSI first decided to limit their activities to two fields, protein:protein interactions and mass spectrometry, whilst also working to establish a single data model that would describe and encompass central aspects of a proteomics experiment. This model will support the various PSI-interchange formats currently under development and also future new areas of interest. Common ontologies will be used across the data model, the interchange formats and ideally, where possible, also will be shared with the micro-array MGED [13] consortium in areas of shared interest such as sample preparation.

The Molecular Interactions group published their Level 1 XML-MI interchange schema at the beginning of 2004 [14] and most of the major molecular interaction databases are already offering data in that format. The Level 1 format only allowed the exchange of protein interaction data. Level 2 expanded this to allow all forms of molecular interaction data to be exchanged and Level 2.5, (Hermjakob et al., in preparation), has further refined the schema to allow the description of situations such as the hierarchical build-up of protein complexes. The success of PSI-MI has lead to the five major public domain databases (BIND [15], DIP [16], IntAct, MINT [17], MIPS [18]) to agree to share curation effort and share data on a nightly basis, a model analogous to the nucleotide database collaboration. The IMEx group hope to initiate data sharing to a common curation standard during 2006 (http://imex.sf.net).

The PSI-MS mzData interchange format is being written to allow both the exchange of experimental data from proteomics experiments involving mass spectrometry and also, with co-operation from instrumentation and search engine manufacturers, to enable researchers to generate data in such a standard directly from their instrumentation. These data written by the vendor data system should be usable directly by search engines, as well as third-party software tools such as spectral databases and other computational tools.

It was agreed that the objectives of this group could best be achieved by aligning with the XML-based standard for analytical information exchange currently being developed by the American Society for Tests and Measures (ASTM) since both standards will have to describe mass spectrometry experiments and results. Standards for spectrometry data such as the ASTM netCDF format and the IUPAC JCAMP format are successful because of broad vendor support and, being computer platform neutral, they have remained readable despite changes in computer technology. As useful as these standards are, it has proved difficult to keep them up to date due to the very rapid changes in mass spectrometry technology. XML was considered the best technology for allowing extensions to keep the standard up to date, while remaining computer platform neutral however the original XML-based format has been switched to Base64 in order to produce more compact files.

The mzData format has now been released (http://psidev.sourceforge.net/ms) and has been implemented by a number of vendors, for example the standard is now supported by

Mascot (Matrix Science), Proteome Systems Ltd, Proteios (Lund University), Phenix (GeneBio), X! Tandem and GPM (Global Proteome Machine Organization). An annual schema release will give sufficient stability for manufacturers whilst still keeping the standard up to date, additional flexibility will be given by a semi-annual CV release. The next minor mzData release will allow XML namespace and versioning. A number of accompanying tools have also been made available, for example mzDataConverter, which allow the user to both convert from mass spectrometry text formats to PSI-MS XML format and mzDataViewer to view and browse stored data in PSI-MS XML format.

The PSI-MS schema is flexible enough to handle a diversity of experiments with a full range of experimental descriptors whilst still remaining compliant with the ASTM model. However, ongoing work will broaden the specification to allow a full description of acquisition and to encompass both mass array and mass intensity. mzData will soon be accompanied by a spectral analysis output format, supporting a common syntax for peptide/protein identification and for protein modification description (AnalysisXML). The mzIdent standard has been designed to capture results from MS search engines and represent the input parameters for analysis algorithms, thus unifying results from different search engines. The requirements for mzIdent include the need to support the identification of both protein and peptides, by accession number or sequence, and must include the ability to describe modifications. Small molecule identification by either CML or SMILES must also be supported. The rapid implementation of such standards by databases such as PRIDE, which is already mzData compatible, and Peptide Atlas [19] will again make data exchange and common curation possible.

Conclusion

Protein interaction data, in particular in combination with protein expression information, is required to add to our understanding of protein function within the cellular environment. Public domain databases can increase their coverage and use to the protein science community by co-operating in data collection and in driving curation standards ever higher. Significant progress has already been made in improving the accessibility and utility of proteomic data, and to date, these efforts have been enthusiastically endorsed by the scientific community. Whilst these efforts are being co-ordinated by the HUPO-PSI, the work is being undertaken by a large body of scientists, representing the worlds of academia, industrial research and instrumentation manufacture and it is to be hoped that they are laying the groundwork for common standards to be widely adopted throughout the entire user community. As these tools and models become more widely available it can be anticipated that they will play a major role in the direction that this important area of biology takes and the eventual utility of the data generated in increasingly high throughput biology.

References

[1]   Mulder, N.J., Apweiler, R., Attwood, T.K.., Bairoch, A., Bateman, A.., Binns, D., Bradley, P., Bork, P., Bucher, P., Cerruti, L., Copley, R., Courcelle, E., Das, U., Durbin, R., Fleischmann, W., Gough, J., Haft, D., Harte, N., Hulo, N., Kahn, D., Kanapin, A., Krestyaninova, M., Lonsdale, D., Lopez, R., Letunic, I., Madera, M., Maslen, J., McDowall, J., Mitchell, A., Nikolskaya, A.N., Orchard, S., Pagni, M., Ponting, C.P., Quevillon, E., Selengut, J., Sigrist, C.J., Silventoinen, V., Studholme, D.J. Vaughan, R. and Wu, C.H.(2005) InterPro, progress and status in 2005. *Nucleic Acids Res.* 2005, 33, 201-205

[2]   Bairoch, A., Apweiler, R., Wu, C.H., Barker, W.C., Boeckmann, B., Ferro, S., Gasteiger, E., Huang, H., Lopez, R., Magrane, M., Martin, M.J., Natale, D.A., O'Donovan, C., Redaschi, N. and Yeh, L.S. (2005) The Universal Protein Resource (UniProt). *Nucleic Acids Res.* 33, 154-159

[3]   Hermjakob, H., Montecchi-Palazzi, L., Lewington, C., Mudali, S., Kerrien, S., Orchard, S., Vingron, M., Roechert, B., Roepstorff, P., Valencia, A., Margalit, H., Armstrong, J., Bairoch, A., Cesareni, G., Sherman, D. and Apweiler, R. (2004) IntAct: an open source molecular interaction database. *Nucleic Acids Res.* 32, D452-455.

[4]   Gavin,A.C., Bosche,M., Krause,R., Grandi,P., Marzioch,M., Bauer,A., Schultz,J., Rick,J.M., Michon,A.M., Cruciat,C.M. *et al.* (2002) Functional organization of the yeast proteome by systematic analysis of protein complexes. *Nature,* 415, 141–147.

[5]   Kanz C, Aldebert P, Althorpe N, Baker W, Baldwin A, Bates K, Browne P, van den Broek A, Castro M, Cochrane G, Duggan K, Eberhardt R, Faruque N, Gamble J, Diez FG, Harte N, Kulikova T, Lin Q, Lombard V, Lopez R, Mancuso R, McHale M, Nardone F, Silventoinen V, Sobhany S, Stoehr P, Tuli MA, Tzouvara K, Vaughan R, Wu D, Zhu W, Apweiler R. (2005) The EMBL Nucleotide Sequence Database. *Nucleic Acids Res.* 33, D29-33

[6]   Brooksbank C, Cameron G, Thornton J. (2005) The European Bioinformatics Institute's data resources: towards systems biology. *Nucleic Acids Res.* 33, D46-53.

[7]   Wheeler,D.L., Chappey,C., Lash,A.E., Leipe,D.D., Madden,T.L., Schuler,G.D., Tatusova,T.A. and Rapp,B.A. (2000) Database resources of the National Center for Biotechnology Information. *Nucleic Acids Res.,* 28, 10–14.

[8]   Harris, M.A., Clark, J., Ireland, A., Lomax, J., Ashburner, M., Foulger, R., Eilbeck, K., Lewis, S., Marshall, B., Mungall, C., Richter, J., Rubin, G.M., Blake, J.A., Bult, C., Dolan, M., Drabkin, H., Eppig, J.T., Hill, D.P., Ni, L., Ringwald, M., Balakrishnan, R., Cherry, J.M., Christie, K.R., Costanzo, M.C., Dwight, S.S., Engel, S., Fisk, D.G., Hirschman, J.E., Hong, E.L., Nash, R.S., Sethuraman, A., Theesfeld, C.L., Botstein, D., Dolinski, K., Feierbach, B., Berardini, T., Mundodi, S., Rhee, S.Y., Apweiler, R., Barrell, D., Camon, E., Dimmer, E., Lee, V., Chisholm, R., Gaudet, P., Kibbe, W., Kishore, R., Schwarz, E.M., Sternberg, P., Gwinn, M., Hannick, L., Wortman, J., Berriman, M., Wood, V., de la Cruz, N., Tonellato, P., Jaiswal, P., Seigfried, T and White, R. (2004) Gene Ontology Consortium. The Gene Ontology (GO) database and informatics resource. *Nucleic Acids Res.* 32, D258-261.

[9]   Garavelli, J.S., *Proteomics.* 2004 Jun;4(6):1527-33.

[10] Martens, L., Hermjakob, H., Jones, P., Adamski, M., Taylor, C., States, D., Gevaert, K., Vandekerckhove, J. and Apweiler, R. (2005) PRIDE: The PRoteomics IDEntifications database *Proteomics* 5, 3537-3545.

[11] Robertson, M. (2004) Reactome: clear view of a starry sky. *Drug Discov Today.* 9, 684-5

[12] Orchard S, Hermjakob H, Taylor CF, Potthast F, Jones P, Zhu W, Julian RK Jr, Apweiler R. (2005) Second proteomics standards initiative spring workshop. *Expert Rev Proteomics.* 2, 287-28

[13] Microarray Gene Expression Data Society.(2004) An open letter on microarray data from the MGED Society. *Int J Syst Evol Microbiol.*, 54, 1917-1918.

[14] Hermjakob, H., Montecchi-Palazzi, L., Bader, G., Wojcik, J., Salwinski, L., Ceol, A., Moore, S., Orchard, S., Sarkans, U., von Meringm C., Roechertm B., Poux, S., Jung, E., Mersch, H., Kersey, P., Lappe, M., Li, Y., Zeng, R., Rana, D., Nikolski, M., Husi, H., Brun, C., Shanker, K., Grant, S.G., Sander, C., Bork, P., Zhu, W., Pandey, A., Brazma, A., Jacq, B., Vidal, M., Sherman, D., Legrain, P., Cesareni, G., Xenarios, I., Eisenberg, D., Steipe, B., Hogue, C. and Apweiler, R. (2004) The HUPO PSI's molecular interaction format--a community standard for the representation of protein interaction data. *Nat Biotechnol.* 22,177-183.

[15] Bader, G.D., Betel, D. and Hogue, C.W.V. (2003) BIND: the Biomolecular Interaction Network Database. *Nucleic Acids Res.* 31, 248-250.

[16] Xenarios, I., Salwinski, L., Duan, X.J., Higney, P., Kim, S. and Eisenberg, D. (2002) DIP: The Database of Interacting Proteins. A research tool for studying cellular networks of protein interactions. *Nucleic Acids Res.* 30, 303-305.

[17] Zanzoni, A. Montecchi-Palazzi, L., Quondam, M., Ausiello, G., Helmer-Citterich, M. and Cesareni, G. (2002) MINT: a Molecular INTeraction database. *FEBS Letts* 513, 135-140.

[18] Pagel, P., Kovac, S., Oesterheld, M., Brauner, B., Dunger-Kaltenbach, I., Frishman, G., Montrone, C., Mark, P., Stumpflen, V., Mewes, H.W., Ruepp, A. and Frishman, D. (2004). The MIPS mammalian protein-protein interaction database. *Bioinformatics.* 2005 Mar;21(6):832-4.

[19] Deutsch EW, Eng JK, Zhang H, King NL, Nesvizhskii AI, Lin B, Lee H, Yi EC, Ossola R, Aebersold R. (2005) Human Plasma PeptideAtlas. *Proteomics.* 5, 3497-3500.

Part B: Genome Annotation Pipelines and Browsers

In: In Silico Genomics and Proteomics   ISBN 1-59454-995-8
Editors: N. Mulder and R. Apweiler, pp. 91-107   © 2006 Nova Science Publishers, Inc.

Chapter 8

Artemis and ACT: DNA Sequence Viewing, Annotation and Comparison Tools

*Tim J. Carver, Stephen Bentley, Nicholas Thomson,
Matthew T.G. Holden and Julian Parkhill*
Wellcome Trust Sanger Institute
Wellcome Trust Genome Campus, Hinxton, Cambridge,
CB10 1SA, UK

Abstract

Artemis is an extensible DNA sequence graphical viewer and annotation tool. [1] The
Artemis Comparison Tool (ACT) uses many of the components of Artemis to highlight
similar regions between two or more nucleotide sequences. Artemis and ACT have been
used to study and annotate a wide variety of prokaryotic (e.g. *Staphylococcus aureus* [2])
and eukaryotic genomes.
Artemis and ACT are written in Java and as such can run on any Java enabled platform,
i.e. UNIX, Macintosh and Windows. Both are free and available for download from the
Sanger Institute web site, http://www.sanger.ac.uk.

1. Historical Background

The origin of Artemis can be traced back to Bart Barrell and colleagues at the Laboratory
of Molecular Biology in Cambridge, UK. In the mid 1980's these pioneers of DNA
sequencing identified the need to write a program to visualise, analyse and annotate the
substantial sequences that they were generating and the task fell to a summer student, James
Crooke. The initial program, named DIANA for DIsplay and ANAlysis of DNA, was written

in FORTRAN for use on the LMB mainframe computer and included a restriction map view, a six frame translated double stranded DNA view and a note field which would ultimately be used to generate an EMBL submission file. There was also a "journal" view for making notes on your findings during annotation for later refinement into a paper.

DIANA proved a great success and was still in use by the Barrell group after they had moved to the Sanger Centre to tackle whole genomes. The entire *Mycobacterium tuberculosis* [3] genome and much of the *Schizosaccharomyces pombe* [4] and *Plasmodium falciparum* [5] genomes were annotated using DIANA but file size limitations and compatibility issues made a re-write necessary. Thus, in 1998, Kim Rutherford wrote the Java-encoded replacement, Artemis (the Greek equivalent of the Roman Goddess of the Hunt, Diana). Artemis is now the program of choice for researchers worldwide interested in sequence analysis.

The arrival of Artemis coincided with an explosion in the availability of whole genome sequences including several closely related bacteria. To facilitate the need to compare such sequences the Artemis code was exploited to produce ACT which allows visualisation of similarities between sequences from individual bases to whole genomes.

2. Installing and Running Artemis and ACT

Artemis and ACT are open source projects and the software is available under a GPL licence. They can be run on many Java enabled platforms, *e.g.* PC windows, Mac OS X, Linux and other UNIX operating systems. The software can be downloaded from the Artemis home page (http://www.sanger.ac.uk/Software/Artemis/) and the ACT home page (http://www.sanger.ac.uk/Software/ACT/). The manuals for each release can also be found on-line.

To get started on a UNIX operating system the software is downloaded and uncompressed. The distributions include shell scripts, called 'art' and 'act', that can be used to run Artemis and ACT respectively.

Windows users can download the applications as jar files. These are recognised as applications that can be run with Java. This means that to launch the Artemis or ACT applications the user simply double clicks on the icon.

Alternatively the applications can be downloaded and run by clicking on the Java Web Start [6] launch links on the home pages. This ensures that the user is running the current version of the software.

The source is available for each stable release from the ftp site. The current development version is available from the public CVS repository (http://cvsweb.sanger.ac.uk/). Code contributions are encouraged to enhance the functionality of Artemis and ACT.

3. Artemis

3.1. Artemis File Formats

Artemis supports several commonly used file formats. A DNA sequence can be loaded into Artemis as a FASTA file. Other sequence formats are supported such as EMBL and Genbank, and the sequence and annotation can be read in from the same file. Alternatively, a feature table can be read in after the sequence has been loaded. Artemis can read feature tables in EMBL, Genbank or GFF formats.

Artemis also reads EMBL or GenBank files that only contain a feature table (*i.e.* FT lines) and no sequence or header lines. These files are called "table" files, or just "tab" files, as they are often named with this as their suffix.

It is possible to overlay a number of different feature tables onto the sequence in Artemis. Each feature table entry can be individually switched on or off in the display so that the display does not get cluttered. Using the one line per entry option, it is also possible for different sources of evidence to be viewed independently and on separate display lines.

3.2. Starting an Artemis Session

When Artemis is run an initial launch window is displayed (Figure 1). It displays the version of Artemis and the genetic code table in use. Sequence files can be opened *via* the "File" menu. Files can be opened using a file selection dialog window ("Open ...") or by using a file manager ("Open File Manager ..."). The file manager can also be used to open sequence files by double clicking on the file name. It can also be used to overlay "table" files onto a sequence opened in Artemis by dragging the file into the main window. The file manager can also be used to carry out other standard file management operations, *e.g.* renaming and deleting files.

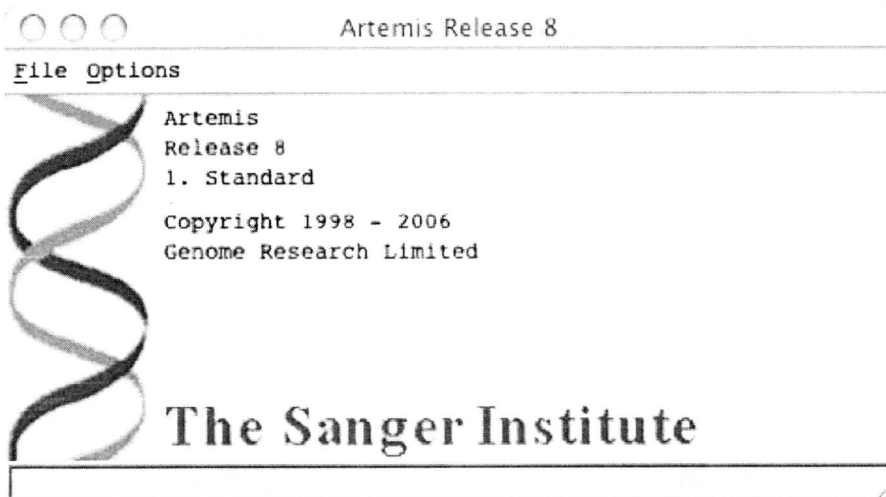

Figure 1. The Artemis Launch Window

Under the "Options" menu (Figure 1), a genetic code table can be selected. The default table is the "Standard Code". This setting affects the display of start and stop codons. When the genetic code table is changed the launch window displays the name of the table being used. The table being used can be changed on the fly *via* this menu and Artemis will automatically update.

3.3. The Graphical Display

A powerful aspect of Artemis is its flexibility in what it can display. There are various views and levels of detail that assist in the annotation process. The "Display" menu in the menu bar at the top of Artemis allows the user to turn on and off various panes in the Artemis window. These panes are described below.

At the top of the Artemis window is a header detailing the number of features selected, see figure 2a. Under this the entry bar that reports the sequence and annotation files that are loaded in, see figure 2b. Several feature tables can be loaded and layered on to the sequence. The default entry is highlighted in yellow. Entries can be made active or inactive by clicking on their checkbox. When an entry is made inactive the features associated with that entry are not displayed. This can be useful as a filtering mechanism to highlight features of particular interest in an entry.

A default entry is specified by the user *via* the "Entries" menu or by right clicking on the entry in the entry bar. This then becomes the entry that any actions become written to. For example when a feature is created then it become part of the default entry. Entries can also be managed using the "Entries" menu. Entries can be renamed, deactivated or removed.

Figure 2d shows the feature pane in which features can be selected, edited, created and deleted. The two grey lines in the middle of this display represent the forward and reverse DNA strands. The 3 lines above the forward strand represent the 3 forward frames of translation and the 3 lines below the reverse strand represent the 3 reverse frames of translation. This is a fairly coarse grained view, displaying in the region of 60 000 bases. The granularity can be increased or reduced using a scrollbar on the right hand side of the pane.

Double clicking on a feature will center the start of the feature in this pane and scroll to this feature and highlight it in the other panes below. Selected feature(s) can be operated on in a variety of ways. For example, Kyte-Doolittle hydophobicity, [7] Hopp-Woods hydrophilicity [8] and coiled coils [9] plots can be shown for the feature by selecting "Show Feature Plots" option from the "View" menu.

If the user right clicks on this pane a pop up menu appears that offers a variety of operations. For example, the stop codons can be ignored in this display by deselecting the option in the 'View' menu.

The display pane shown in figure 2e shows a more detailed view of the sequence. This shows the features at the nucleotide and amino acid level. It shows the six frames of translation. In the middle are the forward and reverse nucleotide strands. The scale shows the forward numbering of the bases.

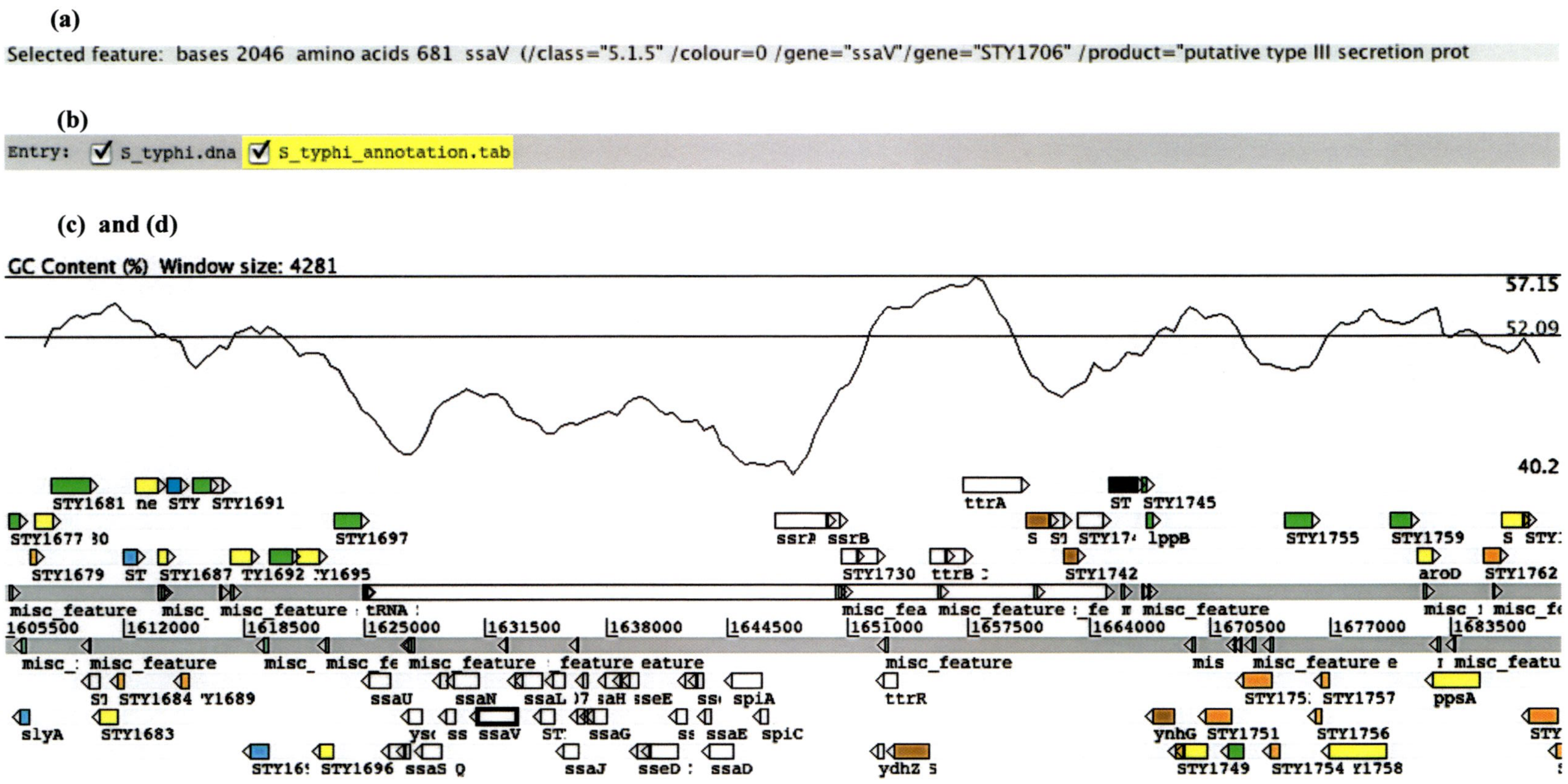

(a)

Selected feature: bases 2046 amino acids 681 ssaV (/class="5.1.5" /colour=0 /gene="ssaV" /gene="STY1706" /product="putative type III secretion prot

(b)

Entry: ☑ s_typh1.dna ☑ s_typh1_annotation.tab

(c) and (d)

Figure 2. A region of sequence from the *Salmonella* Typhi CT18 genome displayed in Artemis. (a) feature selection banner describing the selected feature(s); (b) entry information bar reporting files that are loaded. The default entry is highlighted in yellow; (c) GC content plot; (d) feature display pane showing the gene and other features on one of the 6 frames of translation, or the DNA itself. The forward frames are represented by the 3 top grey bars, the reverse frames by the 3 bottom grey bars, and the DNA by the central two dark grey bars; (Figure continued on next page).

(e)

```
N H I L T T R A Y T E P L L R P D S L T # P * T H C D S R N F I C R * C N R N Q S S A E L E R G G
T H I S * R P A H T L S H C C A L I P S P N H E R I A T P E I L F V D * D V I V T R A R G S A W K E V E
E P Y P * D D P R I H * A T V A P * F P P H L T M E N A L R L Q K F Y L S M M # S # P E L G G A G K R W F
```

```
|1633000        |1633020        |1633040        |1633060        |1633080        |1633100        |1633120        |1633140
AACCATATCCTACGACCCGCCATACACTGAGCCACTGTTGCCGCCCTGATTCCTCACCTTAGTCGATGATGTAATCGTAACCAGACTTTATTTGTCGATGATATCGTAATCGTAACCAGACTGAAAGAGGTGAAC
TTGGTATAGGATGCTGCTGGGCGCGTATGTGACTCGGTGACAACGCGGGACTAAGGAGTGGATTGGTACTTGCCGTAACGCTGAGGTCTTTAAAATAAACAGCTACTACATTAGCATTGGTCTCGAGCCGCCTCGACCTTCTCCAACCTC
```

```
R V V R A Y V S G S N R G S E R V ( + G H V C Q S E L F K I Q R H S T L T W L E A S S L P P
V W I D Q R G A C V S L W Q D A R G E G L Y N A V G S L E K N T S S I T L T V L A R R L S F S T
G Y G S S G R M C Q A V T A G Q N G * R V M F A N R S W F N # K D I I Y D Y G S S P P A P F L H L
```

(f)

```
                    1629738 1631039 c  Similar to Salmonella typhimurium probable secretion system apparatus ATP synthase SsaN SW:SSAN_SALTY (P74857)
CDS                 1631029 1631074 c  Similar to Salmonella typhimurium secretion system apparatus protein SsaV SW:SSAV_SALTY (P74856) (681 aa) fasta
misc_feature        1632574 1632645 c  PS00994 Bacterial export FHIPEP family signature
CDS                 1633059 1633427 c  Similar to Salmonella typhimurium secretion system apparatus protein SsaM   SW:SSAM_SALTY (P74856) (54 aa) fasta
CDS                 1633485 1634477 c  Similar to Salmonella typhimurium secretion system apparatus protein SsaL   SW:SSAL_SALTY () (338 aa) fasta scor
CDS                 1634467 1635141 c  Similar to Salmonella typhimurium secretion system apparatus protein SsaK   SW:SSAK_SALTY () (314 aa) fasta scor
CDS                 1635138 1635686 c  Region similar to Salmonella typhimurium secretion system apparatus protein SsaK SW:SSAK_SALTY (P74853) (314 a
CDS                 1635704 1636453 c  Similar to Salmonella typhimurium secretion system apparatus lipoprotein SsaJ   SW:SSAJ_SALTY () (249 aa) fasta
misc_feature        1636397 1636429 c  Prokaryotic membrane lipoprotein lipid attachment site
CDS                 1636450 1636698 c  Similar to Salmonella typhimurium SsaI TR:Q9ZEF3 (EMBL:AJ224892) (82 aa) fasta scores: E(): 4.8e-29, 95.1% id
CDS                 1636710 1636937 c  Similar to Salmonella typhimurium SsaH TR:Q9ZEF4 (EMBL:AJ224892) (75 aa) fasta scores: E(): 1.2e-29, 98.7% id
CDS                 1636978 1637193 c  Similar to Salmonella typhimurium SsaH or SsaG TR:O30903 (EMBL:AF020808) (71 aa) fasta scores: E(): 7.2e-26, 1(
CDS                 1637287 1637976 c  Similar to Salmonella typhimurium SseG TR:O84952 (EMBL:AF020808) (229 aa) fasta scores: E(): 0, 98.3% id in 22%
CDS                 1637973 1638755 c  Similar to Salmonella typhimurium SseF TR:O84951 (EMBL:AF020808) (260 aa) fasta scores: E(): 0, 96.5% id in 26(
```

Figure 2. A region of sequence from the *Salmonella* Typhi CT18 genome displayed in Artemis. (continued); (e) Base and amino acid pane. This shows the features on the forward and reverse strands in greater detail at the nucleotide level; (f) the lower pane in Artemis lists all the features and their positions and descriptions

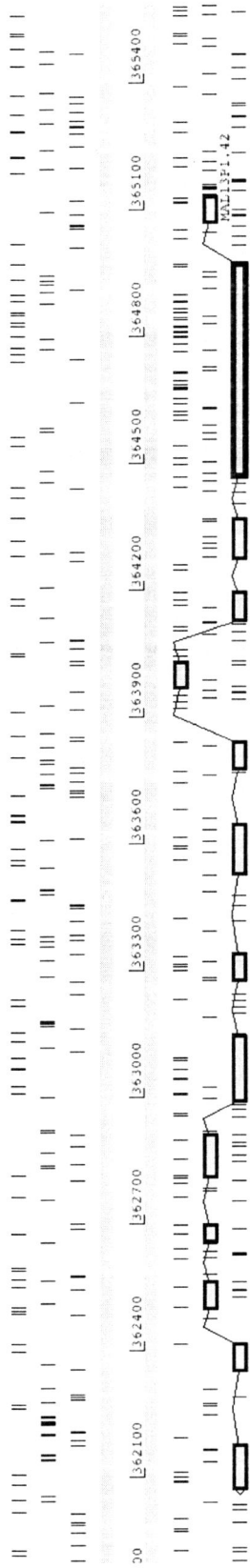

Figure 3. Artemis sequence and feature pane showing splicing

The bottom pane, figure 2f, shows a scrollable list of the features in the sequence. If a feature is selected by double clicking on it in the list, the above displays automatically scroll to that feature. The colour code of a feature is shown on the left-hand side along with the feature type, e.g. coding sequence (CDS) or gene. This is followed by the position on the sequence and if it is on the complementary strand then it is marked with the letter 'c'. Finally the description for the feature is displayed. As with the other panes, right clicking on the pane can access a popup menu. This menu allows the user to adjust what is shown in the feature list.

Figure 2 shows how all the individual sources of information displayed in the multiple Artemis panes can be used to give the user a detailed overview of not only individual CDSs but the whole region. This allows the user to add biological insight to the relatively two dimensional database files, which describe sequence information. The example shown on Figure 2 is a region of sequence from the *Salmonella* Typhi CT18 genome. The G+C plot shows that this region has a lower G+C content than the surrounding DNA. At the border of this anomalous G+C locus is a tRNA gene which is known to act as insertion sites for laterally acquired DNA in bacterial genomes [10]. The annotation associated with each of the genes within this region can be read by selecting the CDS and then viewing the annotation attached to each feature. If the user were to look at the CDSs within this region they would find that they are all associated with pathogenicity, encoding a type III secretion system. [11] This shows one of the key benefits of the Artemis display; many softwares show gene models abstracted from the underlying sequence. The Artemis display does the opposite with the annotation being tied to the sequence. This allows the user to not only see the annotation of individual CDSs but also look at the DNA topology of the region or in fact of the whole genome, thereby adding context (see figure 2). This has the added advantage of allowing the user to challenge the annotation because the sequence is immediately accessible. The user can append or replace annotation or gene models and perform additional analysis that can then be saved within Artemis for future access.

Figure 3 also shows how Artemis is used for both prokaryotic and eukaryotic genome annotation and analysis. The CDS shown is spliced, and has multiple introns. Exons are represented by individual features joined together with a splice symbol (^). In this format it is simple to output the complete spliced or un-spliced sequence of any CDS extracted via the 'view amino acids of selection' option on the 'view' menu. The vertical bars on the 6 translation frames represent the location of stop codons. These are used to indicate potential open reading frames. Start codons can also be shown.

3.4. Navigating and Searching

Features can be selected by clicking on them. Multiple features can be selected by using the SHIFT key when clicking on the features. Alternatively the options under the "Selection" menu can be used. The Artemis Feature Selector is a powerful tool in looking for features based on various criteria, *e.g.* length, position and amino acid motif contained.

Another means of moving around or searching a genome is to use the "Navigator" (under the "Goto" menu). Given a location, feature name, qualifier value, feature key, base pattern or amino acid string the navigator will search for and display the next region that matches the search criteria in the Artemis windows.

3.5. Creating Images (Jpeg/PNG)

Artemis images can be created for the different panels. It is often necessary to print out images from Artemis for publication. To facilitate this a print option has been added (in Artemis 7.0 and ACT 4.0). This can be used to print out the entire Artemis window, without scroll bars, or the panels to be shown can be selected.

3.6. Configuration/Options File

The more advanced Artemis user may wish to change some of the configurable options. Artemis reads these options in when it is launched. It will search 5 different locations on the local file system for the existence of a configuration file. Artemis reads from the locations in the order given below, so the user can override the default options. This is the search order:

- 'etc/options' file in the Artemis distribution directory. This file contains the standard Artemis options and these are read first. Changing this file will change the options settings for all users. On Windows system the standard options are usually contained within the 'clickable' jar file which needs to be unwrapped to be able to alter this file.
- -options command line argument. If the user has specified an options file on the command line with the -options argument it will be read next.
- diani.ini file in the current directory will be read. This is for backward compatibility.
- options, options.txt or options.text file in the current directory will be read.
- '.artemis_options' file in the user's home directory will be read.

The options file can be used to change the appearance and contents of the interface. The options available are given in table 1. The Artemis and ACT distributions each contain an options file that provides sensible default values. However, if required the default values can be changed.

Table 1 Configuration options which can be set in the options files. Defaults are given in square brackets

Option	Description
font_size	Font size for all the Artemis windows. [12]
font_name	Font for all the Artemis windows. [Monospaced]
base_plot_height	The height (in pixels) of the base plots. [150]
feature_plot_height	The height (in pixels) of the feature plots. [160]
draw_feature_borders	Borders will drawn around the features. Also set in the "View" popup menu. ["yes"]
draw_feature_arrows	Direction arrow is drawn at the end of each feature. Also set in the "View" popup menu. ["yes"]
overview_feature_labels	If this option is no then the feature labels in the overview will be off at start up. ["yes"]
overview_one_line_per_entry	If this option is set yes then the overview will start in one line per entry mode. ["no"]
show_list	Feature list is shown on start up. ["yes"]
show_base_view	DNA base view is shown on start up. ["yes"]
features_on_frame_lines	"All Features On Frame Lines" option is set to yes on start up. ["no"]
feature_labels	Feature labels will be shown on start up. ["yes"]
one_line_per_entry	"One Line Per Entry" option is set to yes. ["no"]
genetic_codes	List of the genetic code tables. For each table there is a translation_table_NUMBER entry (NUMBER is its location in the list). [translation_table_1, *i.e.* Standard Code]
translation_table_NUMBER	This is used to lookup codon translations. It must have exactly 64 entries, and there is one entry for each codon.
start_codons_NUMBER	Start codons are defined for each code.
extra_keys	List of keys (separated by spaces) allowed in addition to those specified by EMBL. The official EMBL keys are listed in the feature_keys file.
extra_qualifiers	List of qualifiers (and associated type) allowed in addition to those specified by EMBL. The official EMBL qualifiers and qualifier types are listed in the qualifier_types.
common_keys	List of keys shown in the feature edit window.
undo_levels	Number of undo levels to save (0 to disable). [20]
minimum_orf_size	Minimum size (in amino acid residues) of a "large" open reading frame, controlling which are marked by the "Mark Open Reading Frames" menu option.
dircct_cdit	A value of "yes" will turn direct edit on by default.
feature_dna_programs	List of the possible external programs that can be run on the bases of a feature.
feature_protein_programs	List of the possible external programs that can be run on the translation of a feature.
colour_NUMBER	The colours used to display the feature types. By default there are 18 possible colours. The option names for the colours are colour_0, colour_1, etc.

3.7.UNIX Plugins and Running External Applications

To maximise the potential analyses that Artemis can perform it should be run on a UNIX platform. When run in this way external applications, such as BLAST, [12] FASTA and EMBOSS[13] programs, can be run for features that have been selected in Artemis. The interface executes the external programs by running a script. The script is responsible for running the external program in the required manner. This means that the programs can be sent, for example, to a batch queuing system for processing or simply run in the background.

The programs are split into nucleotide and protein analyses. With the former the selected bases are used and with the later the translated protein sequence is used. When the program has completed the user is notified with a popup window. Any textual output produced can be displayed from the Artemis by selecting the relevant program under "Search Results" under the "View" menu.

Programs can be plugged into artemis by adding their name and a parameter string to the options file. The string can define a database, for example, or may not be used by the script and just ignored. The external program scripts are located in the 'etc' directory of the Artemis directory tree. The convention used for the script name is the prefix 'run_' followed by the program name, *e.g.* run_blastp. For example, the following is used to define the available protein analyses options in Artemis.

```
feature_protein_programs = \
  fasta uniprot \
  fasta uniprot_archaea \
  fasta uniprot_bacteria \
  fasta uniprot_eukaryota \
  fasta uniprot_viruses \
  fasta uniprot_rest \
  fasta malaria \
  blastp uniprot \
  blastp uniprot_archaea \
  blastp uniprot_bacteria \
  blastp uniprot_eukaryota \
  blastp uniprot_viruses \
  blastp uniprot_rest \
  tblastn embl_other \
  pepstats -
```

This makes FASTA and BLAST available for those databases (Uniprot and sub-sections of Uniprot in this case) in the "Run" menu. The last line in the above excerpt from the options file is used to define the EMBOSS application pepstats which calculates protein statistics.

4. Artemis Comparison Tool - ACT

ACT is a comparison tool used to compare two or more sequences. It is used to highlight similar regions between sequences. The display makes use of the same components used by Artemis. This commonality between the two tools has the advantage that it is possible to open an Artemis display from ACT and edit features and the changes will appear in both. It also means that when the user is familiar with one of the tools it becomes straightforward to learn how to use the other.

The sequences and feature tables take the same format as used for Artemis. The only different type of file used by ACT is the file that is used to determine similar regions between the sequences. These regions are usually defined using the output from BLAST. Therefore before running ACT a BLAST comparison file needs to be calculated with one of the sequences acting as the query sequence and the other sequence as a database sequence.

4.1. Running a BLAST Comparison

ACT can read the output of BLAST version 2.2.2 or higher. The BLAST program 'blastn' or 'tblastx' are run to obtain a comparison file for ACT to read. The blastall command must be run with the -m 8 flag, which generates one line of information per high scoring pair. Alternatively MEGABLAST, which is part of the BLAST distribution, can be used to generate a comparison file for ACT. MSPcrunch can also be used. This is a program for UNIX systems that can be used to post-process BLAST version 1 or WUBLAST output into an easier to read format. MSPcrunch must be run with the -d flag.

4.2. ACT Viewer

The query and database sequences used in the BLAST search are represented above and below a central comparison panel (Figure 4). The top sequence is designated the subject sequence, and the lower sequence the query. For each of the sequences the display is as found in Artemis, and can be similarly adjusted as required. The central panel of ACT shows colour bands that are drawn between the regions on the sequence above and below that match. The intensity of the colour represents the percentage identity for the high scoring pairs calculated by BLAST. The more faded regions have a lower percentage identity. The menus are the same as in Artemis (Figure 4) but there is a menu for each of the sequences that are being compared. The sequences can be searched and navigated in exactly the same manner as a sequence in Artemis. Similarly external programs can also be accessed.

Figure 4 shows the comparison of two similar sequences with a region of difference in the centre, The red matches represent the forward matches, reverse matches are coloured blue by default and their boundary is drawn with lines between the matching ends so that they crossover in the middle. The area in the centre of Figure 4 contains a divergent region that has regions that are duplicated and inverted, possibly the result of insertion sequence (IS) element integration and recombination.

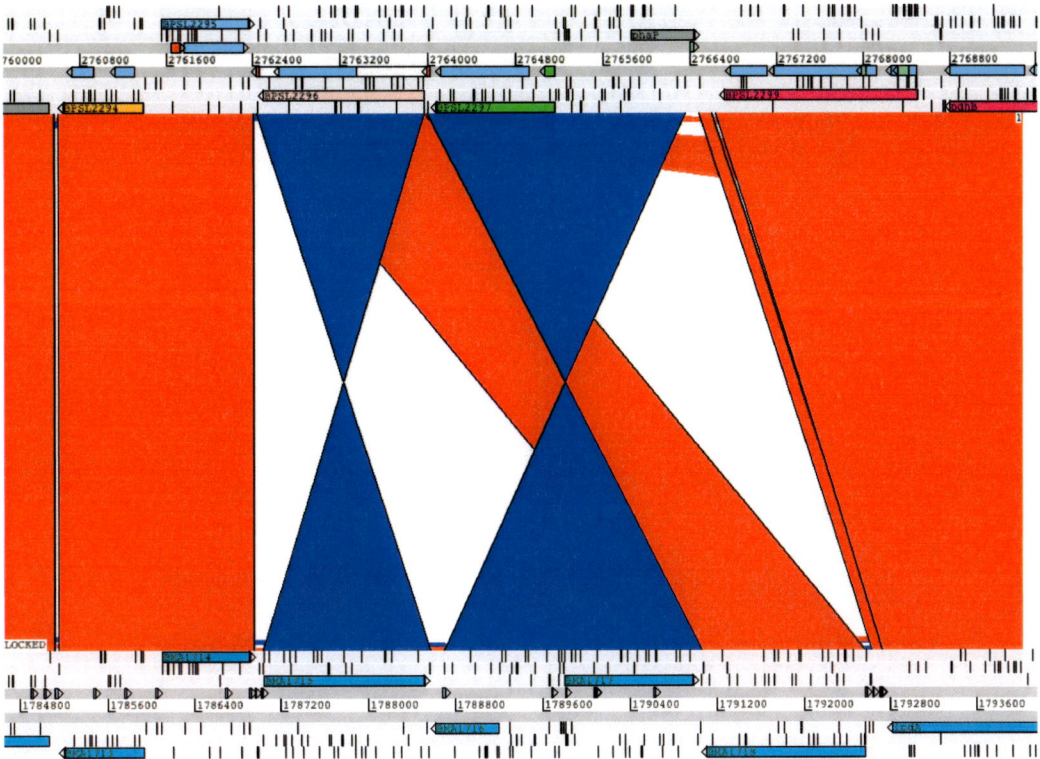

Figure 4. ACT window displaying two sequences and features and the results of a BLASTN comparison.

4.3. Altering the Match View

As with Artemis, the region of sequence viewed in ACT can be altered using the sliders and scrollers in the sequence windows. Double left-clicking on a feature in the sequence windows will centre it in the screen. In addition to the Artemis type menu there is an additional comparison menu (right-clicking on the central panel produces this menu) that contains options for changing the view in the comparison window. Included in the options available is 'Lock Sequences', this locks the subject and query together as they appear in the comparison window, so that if one of the sequences is moved the other will move as well. Conversely, if the sequences are unlocked the sequences will move independently.

To align regions of similarity in the comparison window, it is possible to centre comparison matches by double left-clicking on a match (the comparison match region turns yellow when selected). Where the regions of similarity are distant from one another, it is not very easy to select matches in the comparison window as the match may appear very narrow. To view all the matches that overlap a selected feature or region, the 'View Selected Matches' function from the comparison menu can be used. An additional window appears that lists the matches, displaying the coordinates, score and percentage identity. Selecting a match from the window centres it in the ACT display.

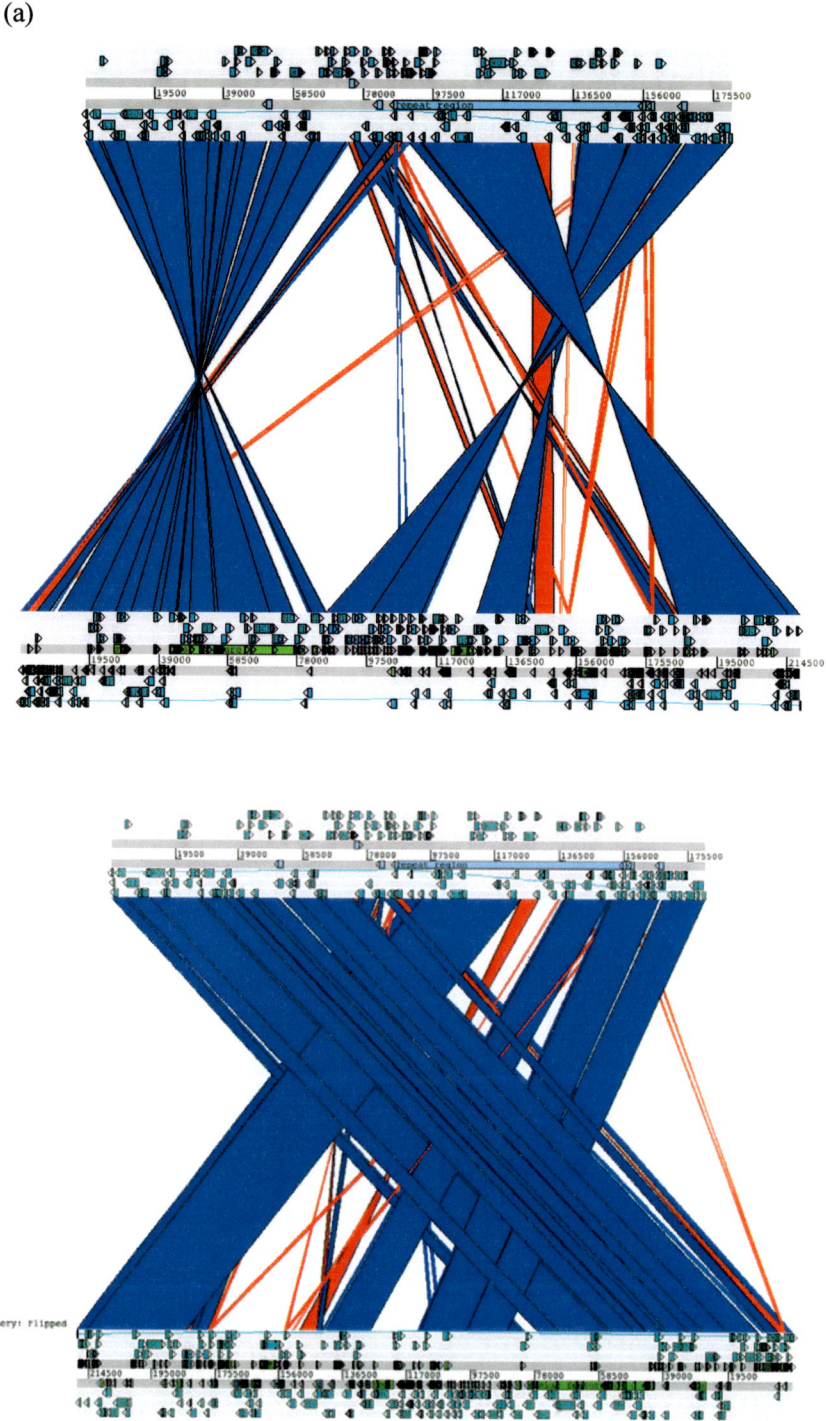

(a)

(b)

Figure 5. Display in ACT of *Salmonella* Typhi plasmids R27 and pHCM1 with BLASTN comparison. The whole plasmid sequences are displayed with the six frame translation (not showing the stop codons). (a) Sequences aligned with the first base of R27 (top; Subject sequence) and pHCM1 (bottom; Query sequence) to the left (b) Query sequence (R27) flipped

For DNA sequences with forward and reverse matches (red and blue), it is possible to alter the orientation of the subject or query sequences in ACT so that match regions can be aligned. This can be a very useful when investigating regions of related DNA that have undergone rearrangements. Figure 5 shows the alignment of two drug resistance plasmids from *Salmonella* Typhi. The two plasmids were isolated more than 20 years apart, however the result from the BLASTN comparison shows that there are regions of DNA shared between the plasmids; pHCM1 shares 169 kb of DNA at greater than 99% sequence identity with R27. Much of the additional DNA in the pHCM1 plasmid appears to have been inserted relative to R27 and encodes antibiotic resistance genes.

The comparison displayed in ACT shows that the two sequences are not co-linear, and that many of the matches are reversed. By selecting the 'Flip Query Sequence' from the comparison menu the orientation of the lower sequence (R27) is flipped (see figure 5b). From the ACT display it is possible to see that there have been several acquisition events that have shaped the evolution of multiple drug resistance in the more modern *S. typhi* plasmid.

With the relentless progress of genome sequencing, more and more related sequences are becoming available. When analysing a sequence, it is often useful to compare it to more than one sequence. In ACT it is possible to view the pairwise comparison of multiple sequences. In Figure 6 the chromosomes of six *Staphylococcus aureus* strains are displayed. Each sequence has been loaded with the appropriate BLASTN comparison. All six chromosomes have conserved gene order along the length of the chromosome, interspersed with small regions of difference. In the case of *S. aureus* much of the sequence divergence is the result horizontal gene transfer.

4.4. Match Filtering

The scrollbar on the right hand side of the comparison view can be used to filter out the smaller regions of match between sequences. The initial setting is a single base pair match between sequences. The minimum match length can be increased by dragging the scrollbar down and only the longer matching regions will be displayed (Figure 6).

Right clicking on the central panel produces a menu that has the option to set cut off BLAST scores and percentage identity scores. These cut off values can be used to highlight high scoring and invariant regions.

Conclusions

Artemis and ACT have proven to be invaluable tools in the annotation process and for DNA viewing. They are very extensible allowing external analyses to be plugged in. Being written in a portable computer language and being freely available make them very accessible to the community.

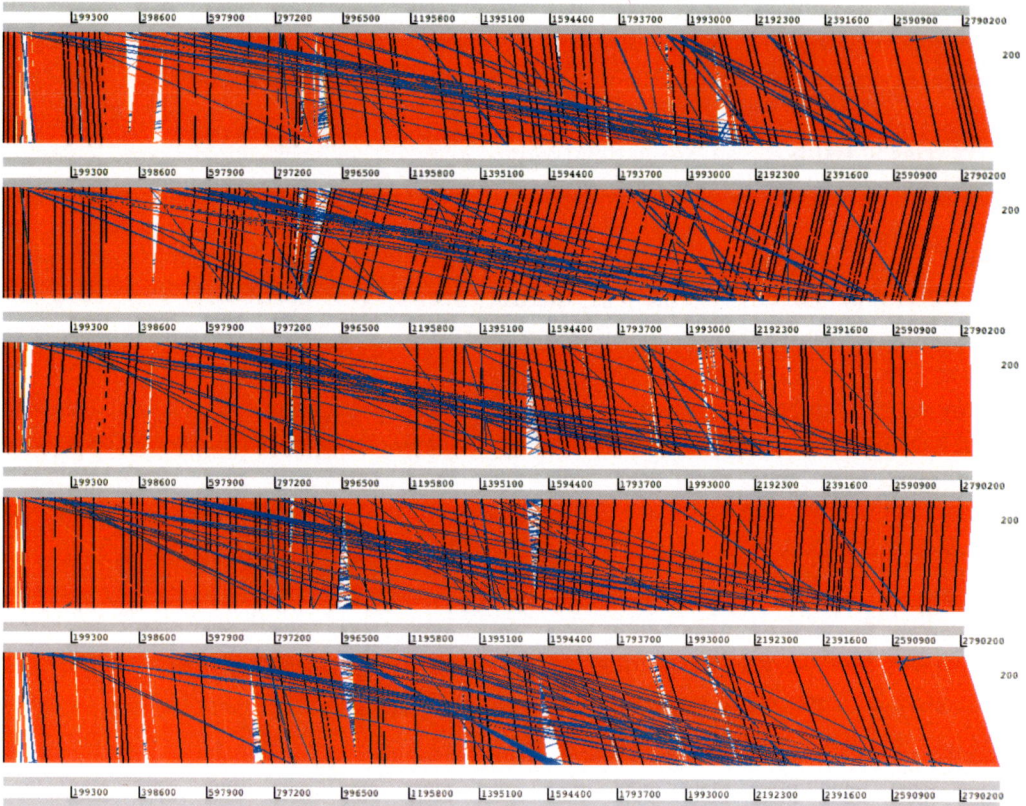

Figure 6. ACT comparison of six chromosomes of Staphylococcus aureus. Pairwise BLASTN comparison of the chromosomes of *S. aureus* strains: COL, Mu50, N315, MW2, MSSA476 and MRSA252 (top to bottom) are displayed with single line per entry, and features removed for clarity. A footprint size cut-off of 200 bp, and a Score cutoff of 100 has been applied to each of the comparison panels

References

[1] Berriman, M. and Rutherford, K. (2003) Viewing and annotating sequence data with Artemis. *Briefings in Bioinformatics*, 4 (2), 124-132.

[2] Holden M.T., Feil E.J., Lindsay J.A., Peacock S.J., Day N.P., Enright M.C., Foster T.J., Moore C.E., Hurst L., Atkin R., Barron A., Bason N., Bentley S.D., Chillingworth C., Chillingworth T., Churcher C., Clark L., Corton C., Cronin A., Doggett J., Dowd L., Feltwell T., Hance Z., Harris B., Hauser H., Holroyd S., Jagels K., James K.D., Lennard N., Line A., Mayes R., Moule S., Mungall K., Ormond D., Quail M.A., Rabbinowitsch E., Rutherford K., Sanders M., Sharp S., Simmonds M., Stevens K., Whitehead S., Barrell B.G., Spratt B.G., Parkhill J. (2004) Complete genomes of two clinical Staphylococcus aureus strains: evidence for the rapid evolution of virulence and drug resistance. *Proc. Natl. Acad. Sci.* 101(26), 9786-9791.

[3] Cole ST, Brosch R, Parkhill J, Garnier T, Churcher C, Harris D, Gordon SV, Eiglmeier K, Gas S, Barry CE 3rd, Tekaia F, Badcock K, Basham D, Brown D, Chillingworth T,

Connor R, Davies R, Devlin K, Feltwell T, Gentles S, Hamlin N, Holroyd S, Hornsby T, Jagels K, Barrell BG, et al. (1998) Deciphering the biology of Mycobacterium tuberculosis from the complete genome sequence, *Nature*, 11;393(6685), 537-44.

[4]   Cole ST, Brosch R, Parkhill J, Garnier T, Churcher C, Harris D, Gordon SV, Eiglmeier K, Gas S, Barry CE 3rd, Tekaia F, Badcock K, Basham D, Brown D, Chillingworth T, Connor R, Davies R, Devlin K, Feltwell T, Gentles S, Hamlin N, Holroyd S, Hornsby T, Jagels K, Barrell BG, et al. (1998) Deciphering the biology of Mycobacterium tuberculosis from the complete genome sequence, *Nature*, 11;393(6685), 537-44.

[5]   Hall N, Pain A, Berriman M, Churcher C, Harris B, Harris D, Mungall K, Bowman S, Atkin R, Baker S, Barron A, Brooks K, Buckee CO, Burrows C, Cherevach I, Chillingworth C, Chillingworth T, Christodoulou Z, Clark L, Clark R, Corton C, Cronin A, Davies R, Davis P, Dear P, Dearden F, Doggett J, Feltwell T, Goble A, Goodhead I, Gwilliam R, Hamlin N, Hance Z, Harper D, Hauser H, Hornsby T, Holroyd S, Horrocks P, Humphray S, Jagels K, James KD, Johnson D, Kerhornou A, Knights A, Konfortov B, Kyes S, Larke N, Lawson D, Lennard N, Line A, Maddison M, McLean J, Mooney P, Moule S, Murphy L, Oliver K, Ormond D, Price C, Quail MA, Rabbinowitsch E, Rajandream MA, Rutter S, Rutherford KM, Sanders M, Simmonds M, Seeger K, Sharp S, Smith R, Squares R, Squares S, Stevens K, Taylor K, Tivey A, Unwin L, Whitehead S, Woodward J, Sulston JE, Craig A, Newbold C, Barrell BG, Sequence of Plasmodium falciparum chromosomes 1, 3-9 and 13, *Nature*, 3;419(6906), 27-31.

[6]   http://java.sun.com/products/javawebstart/

[7]   Kyte J. and Doolittle R.F. (1982) A simple method for displaying the hydropathic character of a protien, *J. Mol. Biol.*, 157, 105.

[8]   Hoop T.P. and Woods K.R. (1981) Prediction of protein antigenic determinants from amino acid sequences, *Proc. Natl. Acad. Sci.*, USA 78:3824, 1981.

[9]   Lupas A, Van Dyke M, Stock J. (1991) Predicting coiled coils from protein sequences. *Science,* 252(5010),1162-4.

[10]   Campbell, A. (2003) Prophage insertion sites Research in Microbiology 154(4) 277-282.

[11]   Watermann., S. R. and Holden., D. W. (2003) Functions and effectors of the Salmonella pathogenicity island 2 type III secretion system. *Cellular Microbiology* 5(8) 501-511.

[12]   Altschul S.F., Gish W., Miller W., Myers E.W. and Lipman D.J. (1990) Basic local alignment search tool. *J Mol Biol* 215, 403-10.

[13]   Rice, P., Longden I. and Bleasby A. (2000) EMBOSS: The European Molecular Biology Open Software Suite, *Trends in Genetics*, 16, 276-277.

In: In Silico Genomics and Proteomics
Editors: N. Mulder and R. Apweiler, pp. 109-123

ISBN 1-59454-995-8
© 2006 Nova Science Publishers, Inc.

Chapter 9

Ensembl's Annotation Pipeline and Its Use in Eukaryotic Genomes

Xosé M. Fernández[1], Steve Searle[2] and Ewan Birney[1]

[1]EMBL Outstation - European Bioinformatics Institute, Wellcome Trust Genome Campus, Hinxton, Cambridge CB10 1SD, UK
[2]Wellcome Trust Sanger Institute Wellcome Trust Genome Campus, Hinxton, Cambridgeshire CB10 1SA, UK

Abstract

This chapter describes the gene annotation process within Ensembl from the raw sequence data to the final gene models clustered into families, SNP information and links to third-party databases delivered in our website (www.ensembl.org). Starting from the gene building system explained in detail (targeted and similarity builds, and pseudogene labelling) to the naming convention. These annotations provide the research community with a powerful framework that can be accessed beyond the website throughout APIs and direct MySQL queries.

Introduction

The whole genetic sequence of human [51],[52],[88] and several animals (including mouse [64], rat [75], chicken [50], a malaria-carrying mosquito (*Anopheles gambiae* [45],[63]) along with a malaria-causing parasite (*Plasmodium falciparum* [38]), plants (*Arabidopsis thaliana* [4] and rice *Oryza sativa* [54]), yeast (*Saccharomyces cerevisiae* [39],[90]) and several other fungi and microbe [36],[21] have been sequenced [59] (with many more species currently being sequenced in centres scattered throughout the world), but this data is little more than raw data, merely strings of nucleotides that in order to be of any use to the scientific community have to be extensively annotated [5],[12],[77]. This annotation is just the beginning of a deciphering process that will starting allow from interpretation of SNPs, leading to the identification of promoter regions or predicting the function of genes.

Gene structures have been analysed since the 1980's, although this analysis was limited to individual genes. The availability of fully sequenced genomes has meant that it is now possible to simultaneously analyse the expression of hundreds of thousands of genes at once, requiring new tools [56] to make sense of this wealth of information.

With many gene already characterised and with genetic maps, the new information had to be integrated [13],[19] enabling the ultimate examination of the genome anatomy, for instance, this new information confirmed that gene density was not uniform along chromosome maps (Figure 1).

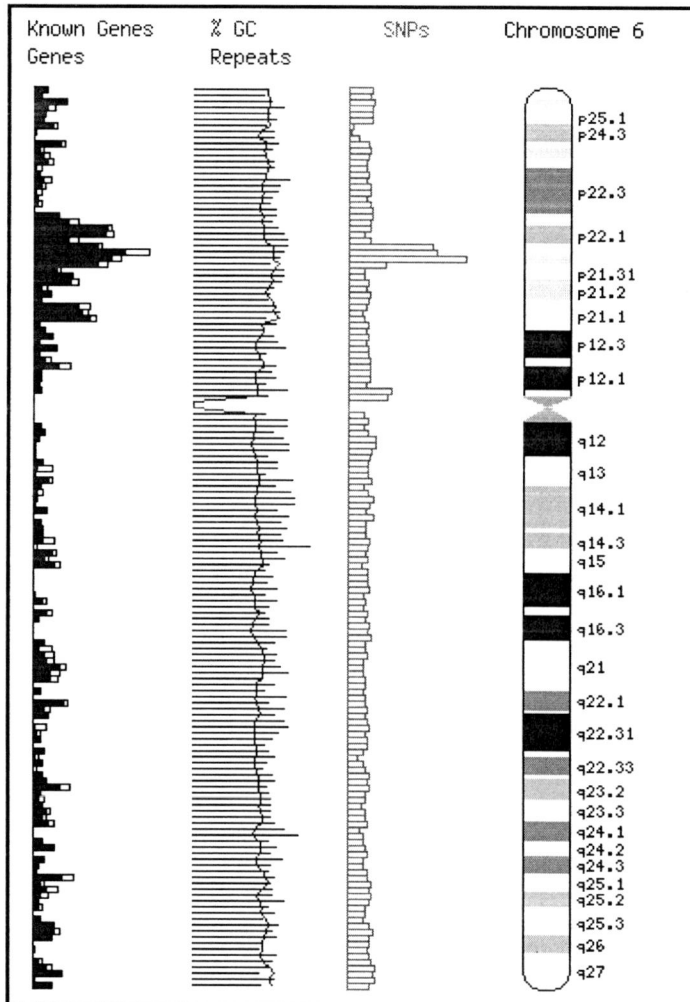

Figure 1. Gene density. Screen shot taken from Ensembl showing how different gene density is across chromosome 6, with a peak around 6p22.2

Ensembl provides a high-level, scalable and integrated framework to annotate and present genetic information from the different genome projects. External annotation can be incorporated into Ensembl by means of the Distributed Annotation System [28]; Ensembl supports sequence annotation editors such as Apollo [60] to read data directly from the databases.

For species where the research community is generating comprehensive manual annotation [6] (mainly human [22],[26],[27],[30],[31],[43],[44],[49],[67] and mouse [53]), Ensembl incorporates those gene and protein sets alongside its own automated annotation [10],[14],[20],[47],[48]. Thus, manual annotation is displayed for some human chromosomes, alongside the Ensembl predictions. For the worm (*Caenorhabditis elegans*) and fly (*Drosophila melanogaster*) genomes, the manually curated genome-wide gene sets are used as imported from their canonical sources (FlyBase [37] and WormBase [41]) in place of an Ensembl set.

Access to all the data produced by Ensembl, and to the software used to analyse and present it, is provided free and without constraints [46],[77] with the rare exception of some proprietary third-party software. Ensembl hosts the following genomes (currently release 33): human, chimpanzee, rhesus macaque, mouse, rat, chicken, dog, cow, opossum, African frog, zebrafish [81], *Takifugu rubripes* [1], *Tetraodon* nigroviridis [55], Anopheles mosquito, fruitfly [68], honeybee, *C. elegans* [85] and baker's yeast.

This chapter will cover the automated sequence annotation pipeline [24],[71] that delivers a consistent genome-wide set of genes, which are displayed in their chromosomal location. The amount of annotation available varies to some extent between species, but Ensembl provides a consistent way of displaying this information so that comparisons of genomic sequence and homologous genes and proteins are facilitated.

Ensembl Gene Building System

The genesis of the Ensembl system was due to the acceleration of the sequencing effort by the International Human Genome Sequencing Consortium in 1999, following the sequencing and publication of the first human chromosome [30]. At that point it was obvious that the only feasible way of annotating the draft sequence in a timely fashion would be automatically and therefore, new algorithms would be required to handle larger and more fragmented genome data rapidly. By 2001 Ensembl was already established, delivering automated annotation for the draft genome assembly [51]. After the announcement of the completion of the Human Genome Project in 2003 some chromosomes have been manually curated and published [22],[26],[27],[30],[31],[43],[44],[49],[67] and can be browsed within Ensembl.

There are a number of *ab initio* gene prediction methods that provide a gene structure prediction solely on the basis of genomic sequence. Most of these algorithms [18] combine local functional signals (such as splice sites or initiation codons), with global statistical properties of protein coding regions. There is a place within the Ensembl annotation system for them, but since even the best methods (e.g. GENSCAN [16]) tend to dramatically over predict genes [18] and miss small exons, other ways had to be explored.

Protein-coding genes (Ensembl gene build is biased towards their prediction) consist of a coding sequence (CDS) usually interrupted by non-coding sequences (introns). Untranslated regions (UTRs) flanking the coding regions can also be interrupted by introns. RNA genes

[32],[33] are also incorporated into Ensembl by means of a set of hand-checked non-coding RNA genes provided by Sean Eddy [61] and Sam Griffiths-Jones (including tRNA, rRNA, miRNA genes amongst others), and Ensembl has recently developed a fully automated system for RNA gene prediction in collaboration with Rfam [40].

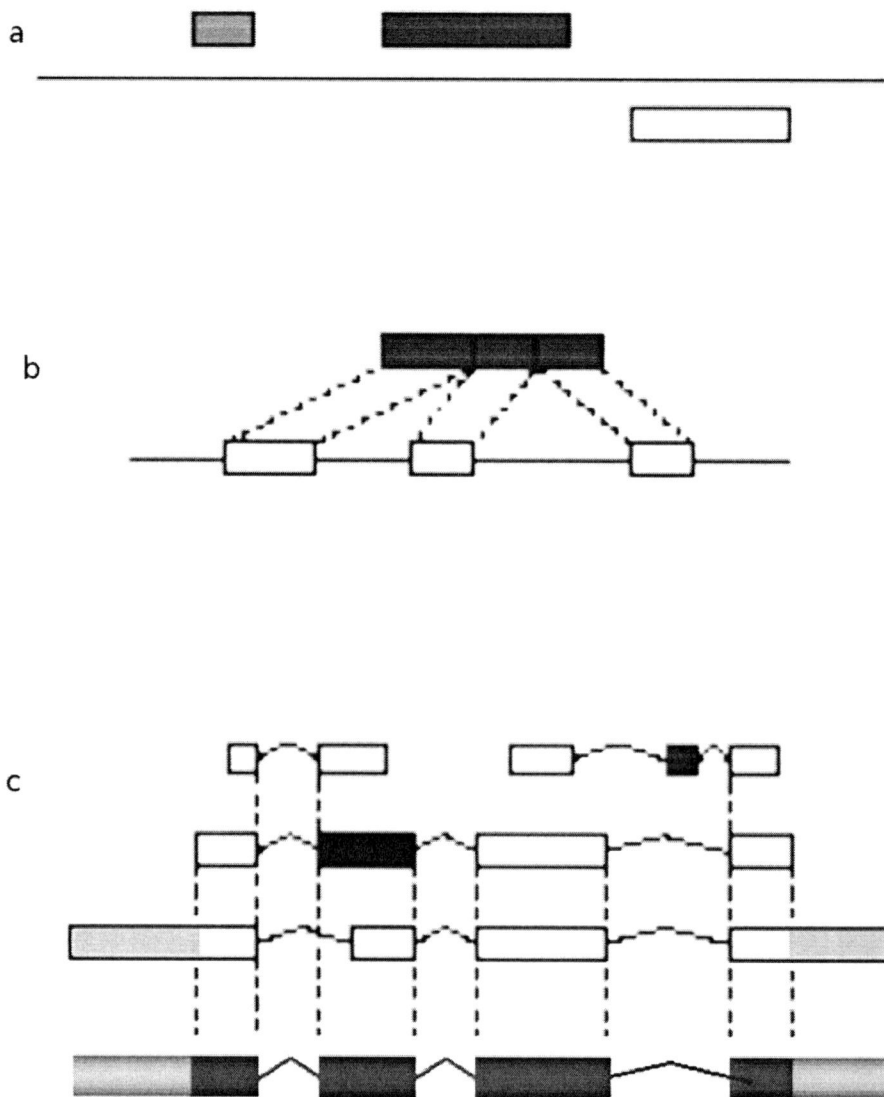

Figure 2. Gene prediction in Ensembl. The prediction process can be briefly described in three stages: (a) alignment of genomics features against the genomic assembly, (b) gene structure prediction and (c) reconciliation of gene predictions from resources providing from different sources (protein databases, expressed sequence tag (EST) and *ab initio* predictions from GENSCAN). Different bioinformatics algorithms are used in this process: (a) BLAST, pmatch are used when the sequence is a protein, whilst BLAST, and exonerate would be used with DNA sequences. (b) GeneWise and Genomescan are used with protein sequences and EST2genome, Genomescan is used with DNA sequences. At the reconciliation stage (c) Genomescan and the Ensembl GeneBuilder are used

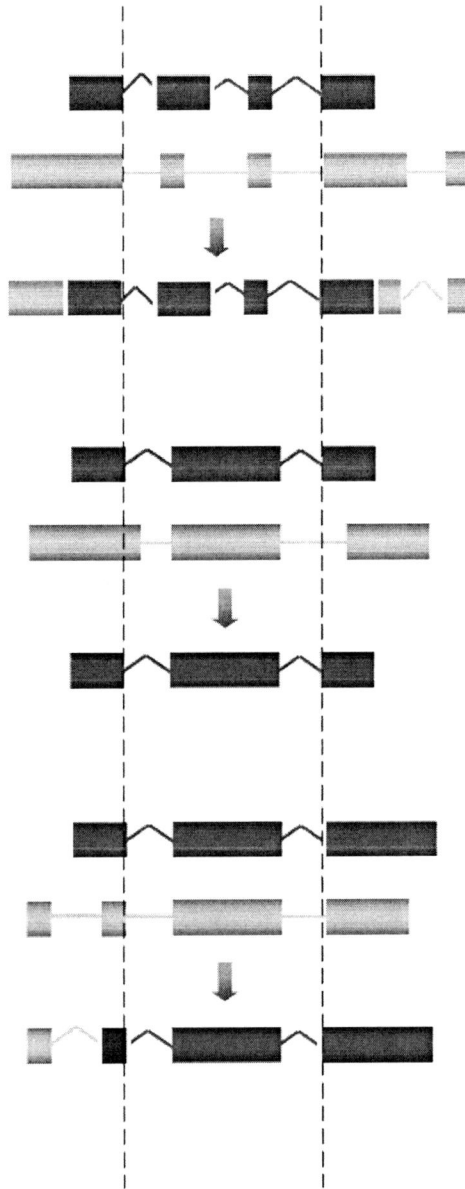

Figure 3. Adding UTRs to GeneWise predictions. a) Ends of exons coincide, thus 1st red exon is extended to include the UTR and the translation start is maintained. Starts for last exons coincide, thus UTR exons are added and the translation stop is maintained (boundary between red and green). For the middle exons, the coordinates of GeneWise-derived (red, splice site model, no UTR, frame information) middle exon are used in preference over exonerates's (no frame, no exon/intron splice model). b) cDNA prediction (green) rejected. Neither the ends of first exons nor the starts of last exons coincide, so the GeneWise-predicted (red) structure is unmodified. c) cDNA prediction with short exons. The ends of the first GeneWise (red) exon and second exonerate (green) exon and the starts of last exons coincide. Even though GeneWise is shorter than exonerate, it is not the first exon of the cDNA prediction and is thus retained. However, the exonerate exon is shorter than GeneWise's and there are no additional exons, so it is rejected.

Table 1. Analyses of the 'Raw Compute'

RepeatMasker	Tool for identifying and masking repeats in genomes. It uses a large database of repeat consensus identified in various eukaryotic species
GENSCAN [16]	Gene-prediction programme that by means of HMMs predicts the presence of a gene in the raw DNA.
DUST	Low-complexity sequences (region of amino acid or nucleotide sequence with a biased residue composition) can result in misleading high scores in sequence similarity searches, as they would reflect compositional bias rather than significant position-by-position alignment. BLASTn queries are filtered with DUST to eliminate these confusing matches from the results. Sequences are treated initially as a heterogeneous mixture with unknown statistical properties, and then attempts may be made to infer these properties. An initial assumption of equal uniform probabilities for the appearance of residues places all possible low-complexity segments on an equal footing. Polypeptide segments rich in common amino acids such as Leu and Ala for instance are treated as no more or no less surprising than segments rich in His, Met, or Try.
BLAST	BLAST [1] ("Basic local alignment search tool") is an algorithm that allows rapid, sensitive similarity searches of protein and nucleotide sequences against Ensembl's annotated organism databases. BLAST heuristic search algorithms look for high-scoring pairs (HSPs) between the query sequence and database sequences. High scoring pairs that satisfy the scoring threshold are returned as local sequence alignments.
CpG	cpgreport (Gos Micklem personal communication) [76] scans a nucleotide sequence for regions with higher than expected frequencies of the dinucleotide CG. These regions are resistant to methylation and tend to be associated with genes which are frequently switched on.
TRF	Tandem Repeats Finder [9] is a program to locate and display tandem repeats (two or more adjacent, approximate copies of a pattern of nucleotides) in DNA sequences.
Fgenesh	Human gene predictor based on pattern recognition of different types of exons, promoters and poly A signals. Based on linear discriminant functions of internal, 5'-coding, and 3'-coding exon recognition, it was designed to find the optimal combination of these and to construct a set of gene models along a given sequence
Eponine	Eponine is a probabilistic method for detecting transcription start sites (TSS) in mammalian genomic sequences, with good specificity and excellent positional accuracy. Eponine models consist of a set of DNA weight matrices recognizing specific sequence motifs. Each of these is associated with a position distribution relative to the TSS.
FistEF	First Exon Finder is a 5' terminal exon and promoter prediction program that employs a set of quadratic discriminant functions recognising features such as CpG islands, promoter regions and first donor sites.

Table 1 Continued

e-PCR [79]	Find matches between STS markers (from UniSTS) and the reference human nucleotide sequence. Run on 1 Mb 'slices' of DNA searching for subsequences that closely match the PCR primers and have the correct order, orientation, and spacing that they could plausibly prime the amplification of a PCR product of the correct length.
tRNAscan-SE [61]	Application for identification of tRNA genes in genomic DNA or RNA sequences, it combines the Cove probabilistic RNA prediction package [32] with the speed and sensitivity of tRNAscan 1.3 [35].
	Ensembl uses tRNAscan and EufindtRNA [70] to identify candidate pol III tRNA promoters as a first-pass; these are subsequently passed to Cove for further analysis, and output if Cove confirms the initial tRNA prediction

RepeatMasker: http://www.repeatmasker.org/
GENSCAN: http://genes.mit.edu/GENSCAN.html   BLAST: http://blast.wustl.edu
TRF: http://tandem.bu.edu/trf/trf.htmlEponine: http://servlet.sanger.ac.uk:8080/eponine/
FirstEF: http://rulai.cshl.org/tools/FirstEF/   ePCR: http://www.ncbi.nlm.nih.gov/genome/sts/epcr.cgi
tRNAscan-SE: http://www.genetics.wustl.edu/eddy/tRNAscan-SE

Briefly, Ensembl combines evidence from various sources in order to deliver a full transcript structure, thus information derived from homology searches can be combined with information from other sources adding a level of complexity to the annotation process. Starting with species-specific proteins and cDNA data, Ensembl creates transcript models placed on the genome assembly; in a later stage data from other species can be added (so that it doesn't take priority over species-specific information), in order to locate additional transcripts. Finally UTRs are added and pseudogenes tagged (Figure 3). An overview of the gene build is provided in Figure 2.

The process of annotation starts with several analyses [71] (termed 'raw compute') which include both analyses required by the gene-build: RepeatMasker (A.F.A. Smit and P. Green, unpublished), GENSCAN [17], dust and BLAST [1], as well as other annotating features such as CpG islands [29], tandem repeats with TRF[9], markers with e-PCR[79] tRNAs with tRNAscan [61], transcription start sites with Eponine and FirstEF [25],[74] (Table 1). Other genome browsers [57] integrate annotation contributed by different groups, but Ensembl in contrast uses this information to deliver a set of predicted gene structures to which additional biological information (expression data, gene ontologies and gene family information) can be added.

For large genomes like human, BLAST analyses could take a long time. In order to speed up this process we BLAST the peptide sequences predicted by GENSCAN against these databases. Thus, we also run GENSCAN over the genomic assembly in order to locate possible genes. These *ab initio* predicted peptides will then be BLASTed against various databases (e.g. protein database, Unigene, etc.) to find regions previously missed.

Once these analyses have been run we can start with the gene build process, the Ensembl analysis and annotation pipeline bases its predictions on experimental evidence which is imported from the manually curated UniProt/Swiss-Prot[3],[7], partially manually curated

RefSeq [73] and automatically annotated UniProt/TrEMBL records [15]. Briefly, proteins are matched to the genome sequence to build raw exon/intron models. UTR regions are extended from the evidence provided by mRNA records from EMBL. The preliminary gene models are refined with GeneWise [11], which contains a splicing model to build exon intron gene models. Broadening the spectrum of its transcript predictions, Ensembl uses evidence from closely related species aiming to reveal new alternative splicing variants.

Pmatch (R. Durbin, unpublished) is a very fast exact matcher (works with either protein or DNA sequences) that looks for identical hits of at least 20 residues, extending them until the exact match ends. Ensembl's 'targetted genebuild' uses this program to place all "known" human proteins (i.e. the human proteome set from UniProt/Swiss-Prot, and the human proteins from RefSeq) onto the genome assembly (at chromosome level), rejecting any matches with less than 25% coverage of the parent protein. We take then the best-in-genome match (based on protein coverage) and any additional matches within 2% of this to the next step with GeneWise. This first initial step reduces the alignment space for each protein from 3 Gb down to about 1 Mb.

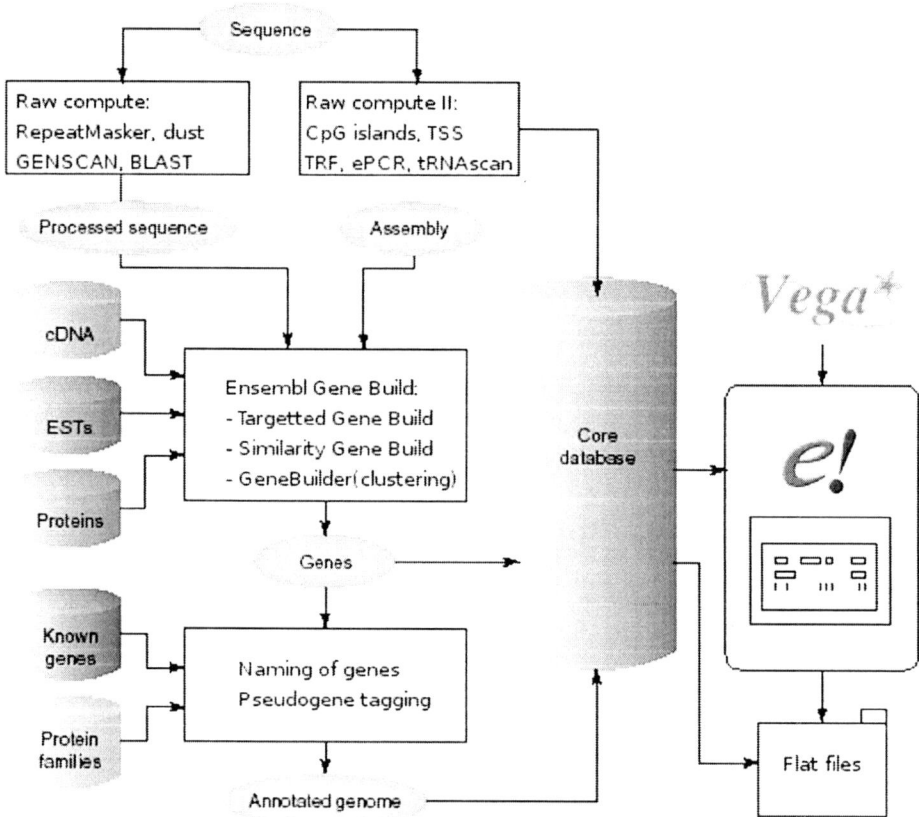

Figure 4. Overview of the Ensembl Automatic Annotation Process. This diagram summarises the Ensembl pipeline, starting with the raw compute, followed by the targetted gene build and similarity gene build delivering the gene predictions which are stored in the core database from where the website runs (key feature of the project, but not the only way of accessing genomic annotation from Ensembl). (See text for more detail)

We run BLAST between the genomic sequence obtained from Pmatch and the protein sequence, extending every BLAST hit with 200 bp on each side to include sequence around the splice sites.

For each protein we use GeneWise which aligns proteins accommodating frameshifts and splice sites, making it ideal for genome annotation. The use of proteins guarantees that the predicted genes will code for protein, the splice site model permits the alignment to 'jump' over introns whilst the fact that it 'allows' frameshifts deals with misassemblies or sequencing errors, making this approach perfect for even draft assemblies. GeneWise is not a fast algorithm, hence the need to reduce our alignment space with Pmatch and BLAST to a size tractable by GeneWise.

This whole process: rough gene positioning with Pmatch, exon location with BLAST and final alignment with GeneWise is delivered in a few hours.

The predicted translation is then compared with the parent protein, rejecting any single-exon predictions showing less than 80% match, whereas for predictions with more exons, we require a match of at least 25% of residues. Another filter implemented takes into account the length of introns (we have observed that long introns are often the first or the last in GeneWise predictions), splitting predicted transcripts when introns exceed 200 kb and the coverage of the parent protein is less than 95%. All these parameters can be modified so that they can be configured differently for different species according to their genomic idiosyncrasy.

Even for the known genes we talk about predictions as the gene structure information is in fact a prediction, although we start off with established sequences. It may not turn out to be the correct one. Since we are often dealing with preliminary or draft assemblies, it may not be possible to align the known sequence 100% onto the genome (e.g. gaps in the assembly). Errors in the sequence assembly may also lead to a well-characterised gene not being placed at all.

Once the known genes are placed, we try to predict additional genes incorporating proteins from other organisms (that will lead to 'novel genes') in what we call the 'similarity build', similar to the 'targetted' stage, but in this case the BLAST analyses are only run on the peptides predicted by GENSCAN and not on the full genomic sequence in order to perform this analysis in a reasonable time. The raw BLAST hits are screened to locate those which don't overlap 'targetted' transcripts, which are subsequently re-BLASTed against the relevant genomic regions; the latter BLAST hit will be fed to GeneWise as before. Thresholds are different here, matches less than 70% of the parent protein will be rejected, as will any prediction with more than 60% of low-complexity sequences in its translation (we use Seg [91] to assess these regions).

At this point, it is worth reiterating that BLAST hits can be used to build more gene models in regions where GeneWise did not build genes. On a final stage, these proteins will be realigned using GeneWise, in case exons were missed.

In a parallel process, we run Exonerate [80] that rapidly builds gapped alignments between cDNAs (from EMBL [83]) and the genome, as it makes a lazy evaluation of the path between HSPs. This approach originally designed for EST genes, is currently being applied to this 'similarity' approach. Exonerate incorporates various models for aligning splice sites.

Thresholds here are set more stringent, 90% coverage and 97% identity is required in order to incorporate those ESTs into our ESTgene build process.

Pseudogenes are sequences of genomic DNA originally derived from functional genes but no longer translated into functional proteins. Pseudogenes are often mis-annotated as functional genes in sequence databases. Recent estimates [42],[87] put the number of human pseudogenes at around 20,000. Pseudogenes are thought to have arisen by two distinct processes:

Unprocessed pseudogenes are considered the result of genome duplication, with a subsequent loss of function of one of the copies due to the accumulation of mutations in the coding or regulatory sequence. *Processed pseudogenes* are thought to have arisen by reverse transcription of processed mRNA, followed by integration into the genome (they lack introns). Pseudogenes are increasingly expected to play an important biological role in eukaryotic genomes [8]. Duplications are believed to be a major source for the formation of new gene expression patterns and functions [72]. Only the latter (processed pseudogenes) are partially annotated in Ensembl (release 33 annotates nearly 2,000 pseudogenes in human). Ensembl does not aim to annotate all pseudogenes, just to label those predictions which have got through the genebuild process but which have characteristics which indicate that they are probably in fact pseudogenes.

Piecing It Together

Ensembl not only produces gene models, but also delivers annotation at the protein level where the predicted proteins are assigned InterPro [66] domains in a process comparable to the 'raw compute'. Two categories can be distinguished at this level, algorithms that will identify InterPro domains (e.g. scanprosite for PROSITE, FingerPRINTScan for PRINTS, pfscan for Pfam [62], etc.), next to other common algorithms (Tmhmm [58], ncoil, Seg and Signal peptide [69]) completing the protein annotation stage.

At this point we reach the moment when we associate Ensembl objects (transcript, translations, etc.) to external external resources (UniProt/SwissProt, RefSeq, EMBL, etc), this mapping finds the best match between Ensembl and the external features.

Amongst other information associated to Ensembl genes, we have variation information (SNPs and information derived from the HapMap project [86]) stored in the variation database where detailed information about linkage disequilibrium (LD) measuring correlation between two alleles (neighbouring genetic variations) in a specific population is stored.

As the number of species sequenced is increasing, comparative analysis is becoming more important within Ensembl, this information is contained in the compara database where gene orthology and paralogy predictions, protein clustering, synteny regions and even whole genome alignments are stored.

Once all these databases are delivered, the final stage would be building BioMart, a query-optimised database to provide flexible, data mining access to Ensembl data.

At the time of writing, Ensembl delivers over 100GB of annotated data (160 GB including multi-species analysis and Mart databases) which can also be directly accessed from the following public MySQL server ensembldb.ensembl.org. All this data has to go through several quality controls, this need is addressed by means of different scripts (healthchecks) starting from the output of the different stages of the pipeline.

With no less that 10 releases a year, all these analyses have to be put together in order to deliver a fully functional website [84] (currently delivering over 5 million hits per week). But Ensembl provides access to the genomes annotated beyond the website, ftp dumps are available from the ftp site[8], where files can be downloaded in flat format or Fasta. Generating these dumps has to be taken into account when planning new releases.

Conclusion and Outlook

A gene-based view of annotated genomes is essential to capitalise on the increase in the sequencing and analysis of model genomes. Ensembl connects maps, sequences, structure, function, and homology data.

Ensembl is involved in a new international collaboration to refine the human gene set, this new effort involves the largest groups currently delivering annotation: from the Sanger Institute, the human and vertebrate analysis and annotation (HAVANA) group (responsible for most of the manually curated Vega database), from the NCBI [89] (the groups involved in the production of automated annotation and RefSeq), the UCSC browser group and UniProt. This productive collaboration aims to resolve transcript sequence differences in order to generate and maintain a set of human genes with stable identifiers, where a consensus was reached between all groups on the CDS part of the gene. This stable gene set is expected to increase progressively as the entire human genome is fully curated. As a by-product of these comparisons, the process is helping both Ensembl and NCBI to improve our automatic gene building methods; in Ensembl, whenever the CDS part of a Vega or Ensembl transcript agrees perfectly with a NCBI transcript and are complete (from ATG to stop codon), these are then propagated through direct transfer into the GeneBuilder stage of the pipeline, and therefore automatically into the next Ensembl gene set, once it was checked that the exon position in those genes remain unchanged in the final new gene set (unlike other genes coming from either the similarity or the targetted build).

References

[1]   Altschul, S. F., Gish, W., Miller, W., Myers, E. W. and Lipman, D. J. (1990) Basic Local Alignment Search Tool. *J. Mol. Biol.* 215, 403-410.

[2]   Aparicio, S. *et al.* (2002) Whole-genome shotgun assembly and analysis of the genome of *Fugu rubripes*. *Science* 297:1301-1310.

[3]   Apweiler R., *et al.* (2004) UniProt: the Universal Protein knowledgebase. *Nucleic Acids Res.* 32, D115-D119.

---

[8]Our ftp server is available at ftp://ftp.ensembl.org/pub/current_human/data; compara
ftp://ftp.ensembl.org/pub/current_multi-species/data/mysql/;and mart
ftp://ftp.ensembl.org/pub/current_mart/data/mysql/

[4] Arabidopsis Genomics Initiative. (2000) Analysis of the genome sequence of the flowering plant *Arabidopsis thaliana*. *Nature* 408, 796-815.

[5] Ashurst, J. L. and Collins, J. E. (2003) Gene annotation: prediction and testing. *Annu. Rev. Genomics Hum. Genet.*, 4, 69-88.

[6] Ashurst, J. L. (2005) The Vertebrate Genome Annotation (Vega) data. *Nucleic Acids Research.* 33, 1:D459-D465.

[7] Bairoch, A. *et al.* (2005) The Universal Protein Resource (UniProt) *Nucl. Acids Res.* 33, D154-D159.

[8] Balakirev, E. S. and Ayala, F. J. (2003) Pseudogenes: are they "junk" or functional DNA? *Annu Rev Genet.* 37, 123-151.

[9] Benson, G. (1999) Tandem repeats finder: A program to Analyze DNA Sequences. *Nucleic Acids Res.* 27, 573-580.

[10] Birney, E. *et al.* (2004) An Overview of Ensembl *Genome Res.* 14, 925-928.

[11] Birney, E., Clamp M. and Durbin R. (2004) GeneWise and Genomewise. *Genome Res.* 14, 988-995.

[12] Birney E., Bateman A., Clamp M. E., Hubbard T. J. (2001) Mining the draft human genome. *Nature* 409, 827-828.

[13] Birney E., Clamp M. and Hubbard T.   (2002) Databases and tools for browsing genomes. *Annu Rev Genomics Hum Genet.* 3, 293-310.

[14] Birney, E. *et al.* (2004) Ensembl 2004. *Nucleic Acids Res* 32, D468-D470.

[15] Boeckmann B., *et al.* (2003) The Swiss-Prot protein knowledgebase and its supplement TrEMBL in 2003. *Nucleic Acids Res.* 31, 365-370.

[16] Burge, C. and Karlin, S. (1997) Prediction of complete gene structures in human genomic DNA. *J. Mol. Biol.* 268, 78-94.

[17] Burge, C. B. and Karlin, S. (1998) Finding the genes in genomic DNA. *Curr. Opin. Struct. Biol.* 8, 346-354.

[18] Burset, M and Guigó, R (1996) Evaluation of gene structure prediction programs. *Genomics.* 34, 353-367.

[19] Cheung,V. G., Nowak, N., Jang, W., Kirsch, I. R., Zhao, S., Chen, X. N., Furey, T. S., Kim, U. J., Kuo, W. L. and Livier, M. (2001) Integration of cytogenetic landmarks into the draft sequence of the human genome. *Nature* 409, 953-958.

[20] Clamp, M. *et al.* (2003) Ensembl 2002: accommodating comparative genomics. *Nucleic Acids Res* 31, 38-42.

[21] Cole, S. *et al.* (1998) Deciphering the biology of *Mycobacterium tuberculosis* from the complete genome sequence. *Nature* 393, 537-544.

[22] Collins, J. E. *et al.* (2003) Reevaluating Human Gene Annotation: A Second-Generation Analysis of Chromosome 22. *Genome Research* 13, 27-36.

[23] Cuff, J. A. *et al.* (2004) The Ensembl Computing Architecture. *Genome Res.* 14, 971-975.

[24] Curwen, V. *et al.* (2004) The Ensembl Automatic Gene Annotation System. *Genome Res.* 14, 942-950.

[25] Davuluri, R. V. Grosse, I. and Zhang M. Q. (2001) Computational identification of promoters and first exons in the human genome. *Nature Genetics* 29, 412-417.

[26] Deloukas, P. *et al.* (2001) The DNA sequence and comparative analysis of human chromosome 20. *Nature* 414, 865-871.

[27] Deloukas, P. *et al.* (2004) The DNA sequence and comparative analysis of human chromosome 10. *Nature* 429, 375-381.

[28] Dowell, R. D., Jokerst, R. M., Day, A., Eddy, S. R. and Stein, L. (2001) The Distributed Annotation System. *BMC Bioinformatics* 2, 7.

[29] Down T. A and Hubbard T. J. P., (2002) Computational Detection and Location of Transcription Start Sites in Mammalian Genomic DNA *Genome Res.*, 12: 458-461.

[30] Dunham, I. *et al.* (1999) The DNA sequence of human chromosome 22. *Nature* 402, 489-495.

[31] Dunham, A *et al.* (2004) The DNA sequence and analysis of human chromosome 13. *Nature* 428, 522-528.

[32] Eddy, S. R. and Durbin, R. (1994) RNA sequence analysis using covariance models. *Nucleic Acids Res.* 22, 2079-2088.

[33] Eddy, S. R. (2001) Non-coding RNA genes and the modern RNA world. *Nat Rev Genet.* 2, 919-929.

[34] Eyras E. *et al.* (2004) ESTGenes: Alternative Splicing From ESTs in Ensembl. *Genome Res.* 14, 976-987.

[35] Fichant GA and Burks C. (1991) Identifying potential tRNA genes in genomic DNA sequences. *J Mol Biol.* 220, 659-671.

[36] Fleischmann, R. D. *et al.* (1995) Whole-genome random sequencing and assembly of *Haemophilus influenzae Rd. Science* 269, 496-512.

[37] Flybase Consortium (2003) The FlyBase database of the Drosophila genome projects and community literature. *Nucleic Acids Res.*, 31, 172-175.

[38] Gardner, M. J. *et al.* (2002) Genome sequence of the human malaria parasite *Plasmodium falciparum. Nature* 419, 498-511.

[39] Goffeau A, *et al.* (1996) Life with 6000 genes. *Science* 274: 546–567.

[40] Griffiths-Jones,S., *et al.* (2003) Rfam: an RNA family database. *Nucleic Acids Res.*, 31, 439–441.

[41] Harris,T.W. *et al.* (2004) WormBase: a multi-species resource for nematode biology and genomics. *Nucleic Acids Res.*, 32, D411-D417.

[42] Harrison, P. M. and Gerstein, M. (2002) Studying genomes through the aeons: protein families, pseudogenes and proteome evolution. *J Mol Biol.* 318, 1155-1174.

[43] Heilig, R. *et al.* (2003) The DNA sequence and analysis of human chromosome 14. *Nature* 421, 601-607.

[44] Hillier L. W. *et al.* (2003) The DNA sequence of human chromosome 7. *Nature* 424, 157-164.

[45] Holt, R. A. *et al.* (2002) The genome sequence of the malaria mosquito Anopheles gambiae. *Science* 298 (5591), 129-49.

[46] Hubbard, T. and Birney, E. (2000) Open annotation offers a democratic solution to genome sequencing. *Nature* 403, 825.

[47] Hubbard, T. *et al.* (2002) The Ensembl genome database project. *Nucleic Acids Res* 30, 38-41.

[48] Hubbard, T. *et al.* (2005) Ensembl 2005. *Nucleic Acids Res.* 33 1:D447-53.

[49] Humphray S. J. *et al.* DNA sequence and analysis of human chromosome 9. *Nature* 429, 369-374.

[50] International Chicken Genome Sequencing Consortium (2004) Sequence and comparative analysis of the chicken genome provide unique perspectives on vertebrate evolution. *Nature* 432, 695-716.

[51] International Human Genome Sequencing Consortium (2001) Initial sequencing and analysis of the human genome. *Nature* 409, 860-921.

[52] International Human Genome Sequencing Consortium (2004) Finishing the euchromatic sequence of the human genome. *Nature* 431, 931-945.

[53] International Mouse Genome Sequencing Consortium (2002) Initial sequencing and comparative analysis of the mouse genome. *Nature* 420, 520-562.

[54] International Rice Genome Sequencing Project, (2005) The map-based sequence of the rice genome. *Nature* 436, 793-800.

[55] Jaillon, O. *et al.* (2004) Genome duplication in the teleost fish Tetraodon nigroviridis reveals the early vertebrate proto-karyotype. *Nature* 431, 946-957.

[56] Kasprzyk, A. *et al.* (2004). EnsMart: A Generic System for Fast and Flexible Access to Biological Data. *Genome Res.* 14,160-169.

[57] Kent, W. J., Sugnet, C. W., Furey, T S., Roskin, K. M., Pringle, T. H., Zahler,A. M. and Haussler, D. (2002) The Human Genome Browser at UCSC. *Genome Res.* 12, 996-1006.

[58] Krogh, A. *et al* (2001) Predicting transmembrane protein topology with a hidden Markov model: application to complete genomes. *J Mol Biol.* 19, 567-580.

[59] Kyripes, N. (1999) Genomes OnLine Database (GOLD): A Monitor of Complete and Ongoing Genome Projects World-wide. *Bioinformatics* 15,773-774.

[60] Lewis, S. E. *et al.* (2002) Apollo: a sequence annotation editor. *Genome Biol.* 3 (12).

[61] Lowe, T. M. and Eddy, S. R. (1997) tRNAscan-SE: a program for improved detection of transfer RNA genes in genomic sequence. *Nucleic Acids Res* 25, 955-964.

[62] Marshall, M. *et al.* (2004) The Pfam Protein Families Database. *Nucleic Acids Res.*, 32, D138-D141.

[63] Mongin, E. *et al.* (2004) The *Anopheles gambiae* genome: An update. *Trends Parasitol.* 20, 49-52.

[64] Mouse Genome Sequencing Consortium. (2002) Initial sequencing and comparative analysis of the mouse genome. *Nature* 420 (6915), 520-62.

[65] Mott, R. (1997) EST_GENOME: a program to align spliced DNA sequences to unspliced genomic DNA. *Comput. Appl.*, 13, 477-478.

[66] Mulder, N. J. *et al.* (2003). The InterPro Database, 2003 brings increased coverage and new features. *Nucl. Acids. Res.*, 31: 315-318.

[67] Mungall, A. J. *et al.* (2003) The DNA sequence and analysis of human chromosome 6. *Nature* 425, 805-811.

[68] Myers, E. W. *et al.* (2000) A whole-genome assembly of Drosophila. *Science* 287, 2196-2204.

[69] Nielsen, H. (1997) A neural network method for identification of prokaryotic and eukaryotic signal peptides and prediction of their cleavage sites. *Int J Neural Syst.* 8, 581-599.

[70]   Pavesi, A. *et al.* (1994) Identification of new eukaryotic tRNA genes in genomic DNA databases by a multistep weight matrix analysis of transcriptional control regions. *Nucleic Acids Res.* 22, 1247-1256.

[71]   Potter, S. C. *et al.* (2004) The Ensembl Analysis Pipeline *Genome Res.* 14, 934-941.

[72]   Prince VE and Pickett FB. (2002) Splitting pairs: the diverging fates of duplicated genes. *Nat Rev Genet.* 3, 827-837.

[73]   Pruitt, K. D., Tatusova, T. and Maglott, D. R. (2003) NCBI Reference Sequence project: update and current status. *Nucleic Acids Res.*, 31, 34-37.

[74]   Ramana, V. *et al.* (2001) Computational identification of promoters and first exons in the human genome. *Nature Genetics* 29, 412-417.

[75]   Rat Genome Sequencing Project Consortium. (2004) Genome sequence of the Brown Norway rat yields insights into mammalian evolution. *Nature* 428 (6982), 493-521.

[76]   Rice, P., Longden, I. and Bleasby, A. (2000) EMBOSS: the European Molecular Biology Open Software Suite. *Trends Genet.*, 16, 276-277.

[77]   Rust, A. G. *et al.* (2002) Genome annotation techniques: new approaches and challenges. *Drug Discov Today.* 1,7 S70-6.

[78]   Salzberg, S., Birney, E., Eddy, S. and White, O. (2003) Unrestricted free access works and must continue. *Nature* 422 (6934), 801.

[79]   Schuler, G. (1997) Sequence mapping by electronic PCR. *Genome Res.* 7, 541-550.

[80]   Slater, G. St. C. and Birney E. (2005) Automated Generation of Heuristics for Biological Sequence Comparison. *Bioinformatics* in press.

[81]   Sprague, J. *et al.* (2003) The Zebrafish Information Network (ZFIN): the zebrafish model organism database. *Nucleic Acids Res.* 31, 241-243.

[82]   Stajich, J. E. *et al.* (2002) The Bioperl toolkit: Perl modules for the life sciences. *Genome Res.* 12 (10), 1611-8.

[83]   Stoesser G. *et al.* (1998) The EMBL nucleotide sequence database. *Nucleic Acids Res.* 26 (1), 8-15.

[84]   Stalker, J *et al.* (2004) The Ensembl Web Site: Mechanics of a Genome Browser. *Genome Res.* 14 (5), 951-955.

[85]   The *C. elegans* Sequencing Consortium. Genome sequence of the nematode *C. elegans*: a platform for investigating biology. *Science* 282, 2012-2018 (1998).

[86]   The International HapMap Project (2003) The International HapMap Consortium. *Nature* 426,789-796.

[87]   Torrents, D. *et al* (2003). A genome-wide survey of human pseudogenes. *Genome Res.* 12, 2559-2567.

[88]   Venter, J. *et al.* (2001) The sequence of the human genome. *Science* 291, 1304-1351.

[89]   Wheeler, D. L. *et al.* (2004) Database resources of the National Center for Biotechnology Information: update. *Nucleic Acids Res.*, 32, D35-D40.

[90]   Wood, V. *et al.*(2001) A Re-annotation of the *Saccharomyces cerevisiae* genome. *Comp Funct Genom* 2,143-154.

[91]   Wootton, J. C. and Federhen, S. (1996) Analysis of compositionally biased regions in sequence databases. *Methods Enzymol.* 266, 554-71.

In: In Silico Genomics and Proteomics
Editors: N. Mulder and R. Apweiler, pp. 125-146
ISBN 1-59454-995-8

Chapter 10

Genome Annotation and Viewing in Vega

Jennifer Harrow and Laurens Wilming

Informatics Department, Wellcome Trust Sanger Institute, Wellcome Trust Genome
Campus, Hinxton, Cambridgeshire CB10 1SA, UK

Abstract

With the increasing amount of large-scale genomic sequencing data, research scientists
are now in a position to look at regions of interest down to the nucleotide level.
However, correctly identifying all the protein-coding loci within the human genome is
still proving difficult due to the limitation of automated prediction procedures, This
chapter describes the different methods of gene prediction, manual assessment,
comparative analysis and experimental verification contributing to the production of a
reference gene sets. It highlights the advantages of manual annotation especially in the
areas of correctly identifying pseudogenes, gene clusters, complex loci, splice variants,
nomenclature and poly(A) features. It also gives an overview of the Vertebrate Genome
Annoation (Vega) Database which is dedicated to the viewing of manual annotation and
has many unique features.

Introduction

Possibly one of the most significant scientific achievements of the 21st century,
sequencing the human genome has led to a rapid expansion of the number of fields trying to
reveal the function of every gene in the various model organisms. But how much closer are
we to finding, and understanding the nature of, every human gene several years after both
public and private efforts published their draft version of human genomic sequence (Lander
et al. 2001, Venter *et al.* 2001)? Initial interpretation of the genomic sequence has relied on
conclusions derived mostly from bioinformatics approaches using *ab initio* predictions,

homology studies and motif analysis to predict gene function. The cataloguing of genes and their regulatory sequences is just the start of a genome-wide approach to discover how the different components interact and contribute to biological processes and physiological complexity. Manual annotation, positioned as it is between the biologist and the informatics communities, plays an essential role in this process, attempting to find the most reliable methods of producing a highly curated gold standard view of the various vertebrate model organism genomes. This chapter describes the Vega browser and the methods of gene prediction, manual assessment, comparative analysis and experimental verification contributing to the production of the gene sets viewable therein.

Vega Database

The Vertebrate Genome Annotation (Vega) system is dedicated to the browsing of manually annotated data. It allows viewing of the manual annotation produced by the Havana group at the Sanger Institute (http://www.sanger.ac.uk/HGP/havana/), as well as the manual annotation from other major sequencing centres, including Genoscope and the Washington University Genome Sequencing Center. At the time of writing about half of all human chromosomes are present in Vega, but we aim to have the complete manual annotation of the human genome available in the course of 2005. Likewise, we plan to be able to present all mouse and zebrafish genome annotation when the sequence is finished quality. Whereas most genome browsers limit themselves to complete genomes, Vega also shows datasets consisting of a sub-region of a chromosome or non-contiguous chromosomal sequences. For example the Del36H region on mouse, representing approximately 20% of chromosome 13 (Mallon *et al.* 2004), is browsable in Vega. In addition, the whole zebrafish (*Danio rerio*) genome assembly can be viewed with individual clones annotated within each chromosome.

In the database that underlies the Vega browser, all gene-related objects (loci, transcripts, exons, translations) are assigned stable, versioned IDs (*e.g.* OTTHUMG00000017411, OTTCANT00000000058, OTTDARE00000028979, OTTMUSP00000000517) very similar to Ensembl IDs, only with ENS replaced with OTT (short for Otter, the name of the database) and a three-letter species identifier for all, including human, IDs. Since the IDs are versioned, when an object is edited the version number increases and the date of the change is saved allowing the user to determine when the annotation was last updated. As will be noticed by users familiar with Ensembl, the Vega browser interface is based on Ensembl (figure 1). Likewise, Otter is a relational database based on an Ensembl schema with an associated client/server system that is able to support interactive updating of annotation (Searle *et al.* 2004). The annotation stored in the Otter backend is either curated directly using Otterlace (a Perl/Tk curation interface working in conjunction with AceDB) or via Otter XML uploads, such as from external groups. Multiple versions of a genome assembly can be stored in the system, with tools to migrate annotation to the latest assembly.

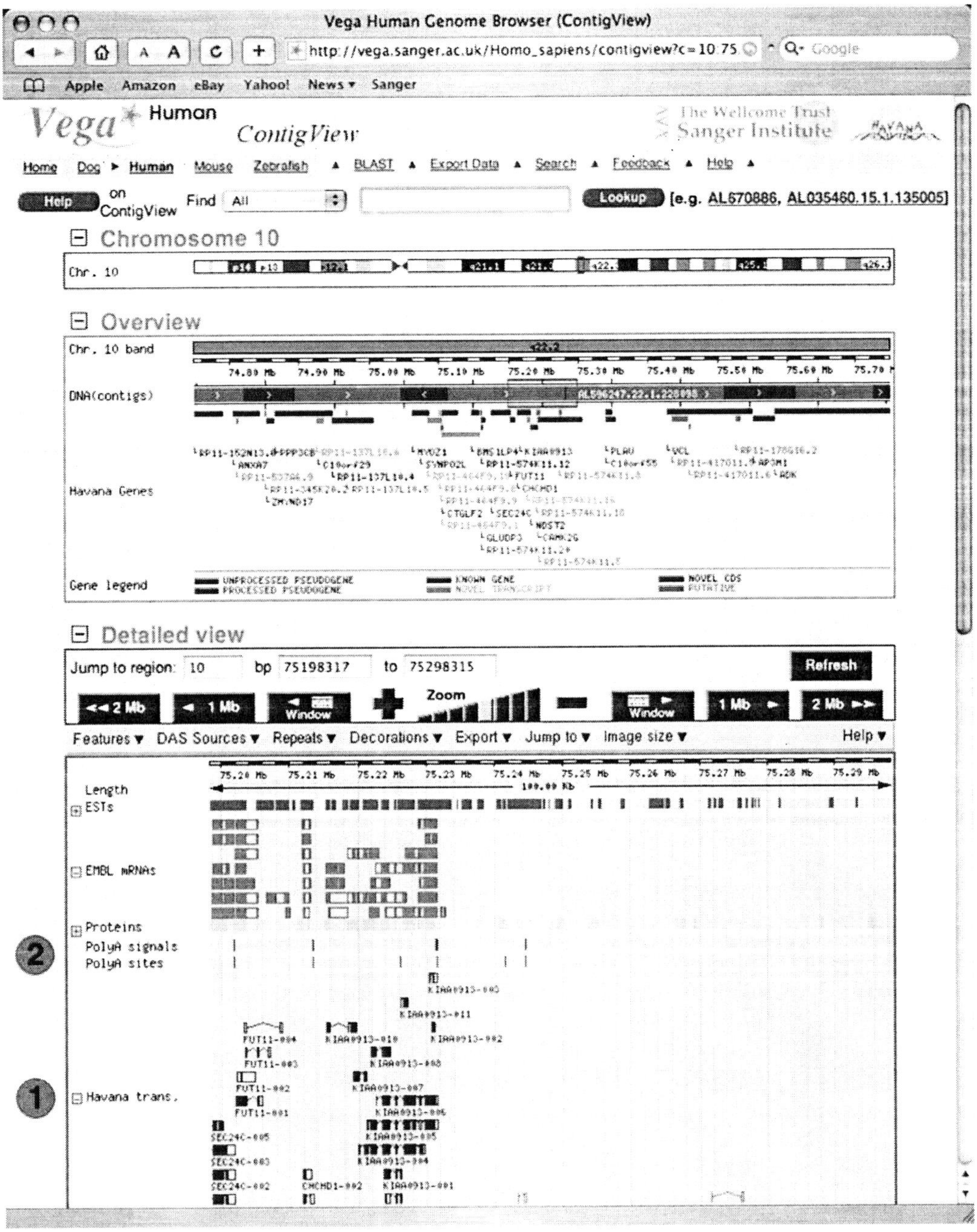

Figure 1a. Vega ContigView of a section of human chromosme 10. The interface is based on Ensembl, but shows only Vega transcripts (1) and has Vega-unique tracks such as polyA features (2).

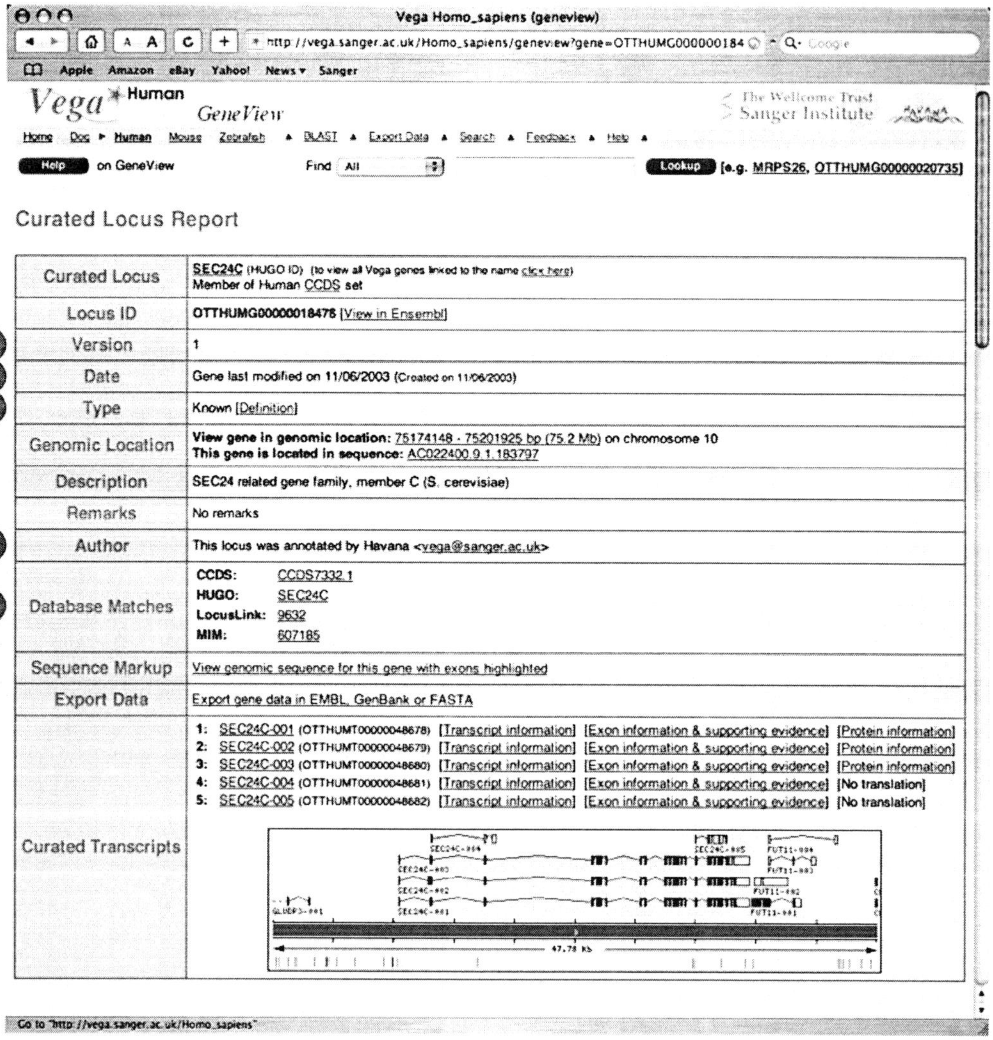

Figure 1b. Vega GeneView of the SEC24C gene, one of the genes from figure 1a. The Ensembl based interface is clear, but there are differences with the equivalent Ensembl view, some of which are marked with a green filled circle. 1: Vega genes are versioned, with the last change date (as well as the creation date) noted (2). 3: classification of the gene as one of Known, Novel CDS, Novel transcript, Putative, Predicted, Pseudogene (Processed, Unprocessed), Ig segment or Ig pseudogene segment. See "Standardisation of annotation and gene classification and nomenclature" section for details. 4: the group responsible for the annotation of the locus. 5: cross-references to the locus in other major databases, including the CCDS database.

Although finished sequence is highly accurate (better than 1 base error in 10000), correction of sequencing and assembly errors and gap filling mean that over large chromosomal regions assemblies are frequently revised. Manual annotation shows one of its strength here, as the detailed examination of multiple sequence alignments (against mRNAs, ESTs and proteins) involved in annotation, allows the detection of potential sequencing errors and aids in unravelling the correct assembly in complex duplicated or highly repetitive regions. Apart from small-scale changes in the assemblies, occasionally large regions are replaced. For example, the human chromosome 6 reference sequence, containing several megabases of sequence representing the MHC region (like the rest of chromosome 6 from a mix of individuals), was replaced with the corresponding sequence from a single haplotype. See "Comparative genome analysis and annotation" below for more details.

Our manual annotation standards require that effectively every annotated exon is supported by homology evidence, though not necessarily with perfect homology (we use imperfect matches to expressed sequences from other family members or other species as long as the exon matches are contiguous and canonically spliced). Vega's evidence viewer shows which mRNAs or ESTs match specific exons so the user can judge the validity of each transcript. This is especially important for splice variants, which generally are built only to the extent of the evidence (mRNA, EST) supporting them.

Vega annotation is used for the consensus CDS (CCDS) project, a joint project by EBI, NCBI, UCSC and WTSI that aims to present a reference standard set of human protein coding genes (http://www.ncbi.nlm.nih.gov/projects/CCDS/).

Automated and Manual Annotation

For the purpose of automated first-pass annotation of the human genome, there was a need for more sophisticated gene prediction tools than were available at the time. Because of the structure of genes in the higher eukaryotes (small exons covering less than 5% of the entire genome in human, large introns, alternative splicing), gene prediction is inherently more difficult for these organisms than for example in the worm *C. elegans*. *Ab initio* gene predictors such as GENSCAN (Burge and Karlin 1997), Fgenes (Solovyev *et al.* 1995), Genie (Reese *et al.* 2000b) and MZEF (Zhang 1997) only use raw sequence data to derive their statistics (*e.g.* codon usage) and signals (*e.g.* splice sites) for gene predictions. When tested against an experimentally validated dataset of mammalian genomic sequences these algorithms suffer from missing, incorrect or over-predicted exons and genes (Rogic *et al.* 2001). Also, these algorithms are limited to finding protein coding regions, leaving out the 3' and 5' UTRs. By adding protein or mRNA sequence similarity data, Genewise (Birney and Durbin 2000) and GenomeScan (Yeh *et al.* 2001) greatly improve on *ab initio* predictions though gaps in their predictions remain (Reese *et al.* 2000a). However, for first-pass annotation of complete or draft genomic sequence, this combined approach, as used by Ensembl for the human genome sequencing consortium effort (and later for many other genomes) (Hubbard *et al.* 2002) and by Otto (developed by Celera) (Venter *et al.* 2001), is sufficient to locate most genes even if the structure is not necessarily one hundred percent accurate.

As more genomes reach finished quality, attention is now turning to obtaining a "gold standard" gene set for the human, mouse and other genome sequences and manual curation is currently the only realistic option for attaining this goal. In the following sections we discuss the various issues relating to gene annotation and the role of Vega's manual annotation in these.

Standardisation of Annotation and Gene Classification and Nomenclature

In order to standardize definitions of gene features between the different research groups that are performing manual annotation of genomic sequence, collaborators involved in submitting annotation data to Vega have attended a series of human annotation workshops (HAWK) (http://www.sanger.ac.uk/HGP/havana/hawk.shtml). These were used to define a standard of annotation (see http://www.sanger.ac.uk/HGP/havana/docs/guidelines.pdf) subsequently used for producing the data in Vega. A common factor is that all annotated gene structures must be supported by transcriptional evidence, either from cDNA, EST or protein sequences. The following are the gene types first used in human chromosome 20 annotation (Deloukas *et al.* 2001) and later adopted as the Vega standard:

-   *known genes* are identical to human cDNA or protein sequences identified by a GeneID in Entrez Gene (http://www.ncbi.nlm.nih.gov/entrez/query.fcgi?db=gene); this criterium is used at the time of annotation. Naturally, as time passes, genes would move to this category.
-   *novel genes* have an open reading frame (ORF) and are identical, or have homology, to cDNAs or proteins but do not fall in the above category; these can be known in the sense that there are mRNA sequences for them in the public databases, but they are not yet represented in Entrez Gene or have not received an official gene name. They can also be novel in that they are not yet represented by an mRNA sequence in the species concerned.
-   *novel transcripts* are as above but no ORF can be unambiguously assigned; these can be genuine non-coding genes or they can be partial genes because of the limits of the evidence they are based on.
-   *putative genes* are identical, or have homology, to spliced ESTs but are devoid of a significant ORF; these are generally short two or three exon genes or gene fragments.
-   *pseudogenes* have homology to proteins but generally suffer from a disrupted CDS and an active gene can normally be found at another locus; these can be processed or unprocessed and sometimes even have an intact CDS or an open but truncated ORF, in which case there is other evidence used (for example genomic poly(A) stretches at the 3' end) to classify them as pseudo.

For the annotation of chromosome 14, Genoscope used an additional classification called "predicted genes" to describe genes based on *ab initio* predictions that have at least one exon supported by biological or similarity data (unspliced ESTs, mouse or Tetraodon genomes or expression data from Rosetta) (Heilig *et al.* 2003). Together with putative genes these genes serve as targets for experimental validation (Ashurst and Collins 2003). Other special tags have been used for immunoglobulin segments and immunoglobulin pseudogenes on human chromosomes 22 (Dunham *et al.* 1999) and 14 (Heilig *et al.* 2003).

Using correct gene nomenclature is important for maintaining consistency in an annotation database, especially when working with syntenic regions across species or haplotypes within a single species. There is therefore close interaction between annotation staff involved in the Vega project and the human (HGNC) (Wain *et al.* 2002), mouse (MGD) and zebrafish (ZFIN) nomenclature committees. At the time of initial annotation, lists of annotated protein sequences are exchanged and up-to-date gene/locus symbols provided (and new symbols assigned where necessary and possible). If an approved symbol is not available for a gene locus, an interim identifier is used which is usually in the format clonename.number, *e.g.* RP11-694B14.5. Ongoing communication ensures that the ever-changing gene symbols are continuously updated.

Alternative Splicing

One way complex organisms such as higher vertebrates are thought to have increased proteome and transcriptome complexity from estimated gene numbers that are not that different from those of much simpler organisms such as worm *C. elegans* and fruitfly *D. melanogaster*, is alternative splicing of pre-mRNA (Graveley 2001). It provides a versatile mechanism by which major developmental functions can be controlled and transcripts can be expressed in a cell or tissue specific manner (Lopez 1998, Yeo *et al.* 2004, Taneri *et al.* 2004, Anderson *et al.* 2005). Alternative 5' exons enable the use of different promoters for different tissues (Anderson *et al.* 2005), while exon skipping, alternative exons and alternative splice sites change the actual protein translated from the mRNA, and therefore very likely its function (Taneri *et al.* 2004). Analysis of alternative transcripts from the annotation of finished human chromosomes shows 59% of chromosome 22 (Dunham *et al.* 1999), 35% of chromosome 20 (Deloukas *et al.* 2001) and 54% of chromosome 14 (Heilig *et al.* 2003) gene loci display alternative splicing, with an average 2.5 alternative transcripts per locus. This is in line with estimates of alternative splicing in the human genome as a whole of around 50% of genes (Lander *et al.* 2001, Venter *et al.* 2001). The most extreme case of alternative splicing is seen in the *Drosophila Dscam* (Down syndrome cell adhesion molecule) gene which in theory can generate over 38000 distinct mRNA transcripts (Schmucker *et al.* 2000). In man, this gene is located on chromosome 21q22 and is thought to be involved in neural differentiation and contribute to the central and peripheral nervous system defects in Down Syndrome (Yamakawa *et al.* 1998).

There have been suggestions that EST coverage, rather than complexity of organism, influences the level of alternative splicing detection. After examining seven different eukaryotic organisms with sufficient EST and mRNA coverage, Brett *et al.* found that alternative splicing can be detected in a large number of organisms, including invertebrates (Brett *et al.* 2002). Similarly, when analysing exon skipping events in human chromosome 22, it was observed that high EST coverage influences the number of alternative transcripts identified (Hide *et al.* 2001). Concerns have also been raised about the reliability of some of the EST data leading to incorrect predictions using automated gene finding (Modrek and Lee 2002). For example, looking at genomically aligned ESTs it is clear that some ESTs were oligo(dT)-primed from genomic contamination or represent apparently incompletely spliced heteronuclear RNA (hnRNA). However, the latter may not be artefacts but may act as translation regulators through the early translational termination caused by the variation

(Smith and Valcarcel 2000, Modrek and Lee 2002). Examination of 16162 mRNAs from Locuslink (Pruitt and Maglott 2001), identified 1106 loci that generate 1989 distinct alternative transcripts that are targets for nonsense mediated decay (NMD) (Lewis *et al.* 2002). Therefore, the contribution of alternative splicing to proteome diversity may be balanced by a yet unappreciated regulatory role in gene expression (Stamm *et al.* 2005).

Kapranov *et al.* (2002), Bertone *et al.* (2004) and Schadt *et al.* (2004) suggest that there is still a considerable number of potential splice variants undiscovered. Probing of genome tiling arrays for transcribed regions and aligning these regions against annotation, highlighted many potentially transcribed regions in introns, suggesting un-annotated alternative exons. Mapping SAGE tags to the human genome, Saha *et al.* also found many matches to introns (Saha *et al.* 2002). Some intronic matches can of course correspond to novel genes located in introns of other genes and since the methods used generally do not distinguish between one strand or the other, the intronic matches could also represent novel genes on the opposite strand.

One of the contributions of manual annotation to genome analysis is the careful annotation of splice variants. Our criteria for annotation of alternative transcripts is that EST/cDNA evidence used for gene annotation should splice. This way we exclude those transcripts that appear to have novel introns (mostly in the 3' UTR) that are probably artefacts caused by recombination between identical sequences flanking the deletion. A CDS is not a requirement for splice variants: as mentioned above, non-coding transcripts can have a regulatory function. In fact we find that close to half of the variant transcripts we annotate do not appear to have a CDS. ESTs from different species are used as evidence to predict different transcripts (on the condition that canonical splice sites are used and the alignment is ungapped) on the premise that they may supply a wealth of information from different developmental stages not available for the species being studied. Genome comparison studies between mouse and human have generally observed that gene structures are conserved (Batzoglou *et al.* 2000, Kan *et al.* 2002) and that at least some alternative splicing events are as well (Yeo *et al.* 2005).

Poly(A) Sites and Signals

Determining the 3' end of a gene is relatively straightforward because cDNA synthesis usually starts from oligo-d(T) primers hybridized to the poly(A) tail at the 3' of the mRNA (though naturally these primers also bind to internal poly(A) stretches found in some genes). Therefore there is a strong 3' bias amongst EST sequences, which can be used to an advantage for finding the poly(A) features of a gene. ESTs also clearly mark out alternative ends of a gene. Alternative polyadenylation is found to take place in a large number of higher eukaryotes, presumably as a regulatory mechanism because it occurs in a tissue dependent manner, and has been implicated in disease (Edwalds-Gilbert *et al.* 1997).

Beaudoing *et al.* (2000) found, by comparing human 3' UTRs from UTRdb (Pesole *et al.* 2002) to ESTs, ten different polyadenylation signals but the vast majority represented by only two types: AAUAAA (58%) and AUUAAA (15%). The signal is almost always found in the region around 16 nucleotides upstream of the poly(A) site. AAUAAA seems to be processed more efficiently than the others and in cases of alternative polyadenylation more prevalent at the 3' most site. This indicates that the major form of the alternatively polyadenylated mRNAs will in general be the longest and that the longer forms are more stable than the corresponding shorter forms.

Whereas manually determining poly(A) features is fairly easy, automating this process is less straightforward, especially when it comes to finding alternative poly(A) sites. The latter is mainly caused by the fact that the poly(A) signals are only six bases long and are therefore very abundant in 3' exons (and the rest of the genome for that matter). However, prediction programs such as POLYADQ exist which can predict both AAUAAA and AUUAAA-dependent poly(A) sites in the human genome (Tabaska and Zhang 1999).

Our manual annotation criteria require a run of at least four non-genomic A residues at the end of an aligned cDNA or EST sequence before we consider the 3' end of a gene established. Signals, in most cases detected within 60 bases of the site, are only annotated in conjunction with sites, though sites can be annotated in the absence of signals.

Pseudogenes

A pseudogene is defined as a non-functional copy of a gene. Pseudogenes are generated by two different mechanisms: retrotransposition or duplication of genomic DNA. What most pseudogenes of both types have in common is frameshifts and/or in-frame stop codons in the coding region, presumably rendering them non-functional. But where they differ is in their structure. Pseudogenes arising from retrotransposition are known as processed pseudogenes and are generated by insertion into the genome of double stranded sequence formed by reverse transcription of single stranded processed (*i.e.* intron-less) mRNA (Vanin 1985). Processed pseudogenes lack introns and 5' promoter sequences and some times even a smaller or larger part of the 5' end of the original gene. Often a genomic poly(A) tract or A-rich region is present at the 3' end, marking the downstream insertion point. In contrast, pseudogenes arising from gene duplication, termed unprocessed, have an exon structure similar to their ancestral gene. Duplication of DNA segments is essential for the development of complex genomes, yet exactly how this occurs is still under debate (Cooke *et al.* 1997, Ganfornina and Sanchez 1999). Because of the different mechanisms generating the two types, they occur in different patterns: processed pseudogenes can be found throughout the genome, often inside other genes, without any positional relationship with their ancestral genes, whereas unprocessed pseudogenes are generally found close to, and often clustered with, one or more active copies of the gene. In some clusters over half of the genes are pseudogenes (V1rh +V1ri vomeronasal receptor gene cluster on mouse chromosome 13 (Mallon *et al.* 2004)) and sometimes all the copies have degenerated into pseudogenes (vomeronasal receptor gene cluster on human chromosome 6 (Mallon *et al.* 2004)).

Why are pseudogenes important, since most are not expressed anyway? Firstly, some are involved in disease identification. For example, ribosomal proteins (RP) are implicated in human genetic diseases such as Noonan Syndrome (Kenmochi *et al.* 2000). Therefore characterising all pseudogenes in the genome will help researchers design specific primers for the functional RP genes. Secondly, pseudogenes, combined with orthology and homology information, can be used to study evolution and phylogenetic relationships between different organisms. Finally, processed pseudogenes can be used to study retrotransposition, as they can be seen as a kind of repetitive element similar to SINEs (many of which are sometimes considered tRNA pseudogenes (Weiner *et al.* 1986, Daniels and Deininger 1985)). Because pseudogenes are more diverse and complex than repetitive elements in terms of sequence length and GC content it makes them a useful tool to study the evolution and dynamics of genomes.

The categorisation of ORFs as functional genes or pseudogenes is made difficult by the fact that some processed pseudogenes still have an intact CDS and may have maintained or (re)gained function (*e.g.* human testis specific PGK gene (McCarrey and Thomas 1987)). Also, some unprocessed pseudogenes have a CDS very similar to their expressed versions with only a slightly truncated 3' end due to an in-frame stop codon close to the regular stop codon or a slightly extended or truncated 3' end due to a frameshift close to the regular translation terminus. Moreover, some processed or unprocessed genes have a CDS of virtually the same length as the original, but of (slightly) different composition owing to two frameshifts that cancel each other out, *i.e.* the sequence between the frameshifts is translated from a different frame (relative to the original gene) and drops back after the second frameshift. In all these cases automated annotation tools would consider these genes as expressed genes. Conversely, expressed pseudogenes that function as non-coding RNAs (for example Makorin1-p1 (Hirotsune *et al.* 2003)) may escape automatic annotation. Manual annotation allows us to judge each case on its merits and take other information into account, such as from publications. Whereas these types of pseudogenes may never be correctly typed through an automated approach, with a set of well thought out parameters it is possible to detect a large proportion of pseudogenes. For example, Harrison *et al.* (2002) set out to identify and analyse pseudogenes on human chromosomes 21 and 22 (Hattori *et al.* 2000, Dunham *et al.* 1999), using a method that searches for DNA regions similar to known proteins but with obvious frameshifts or in-frame stop codons, minimal overlap with annotated known genes and, optionally, evidence of a 3' poly(A) tract. They found a considerable number of new pseudogenes not reported by the respective sequencing centres.

In conclusion, where automated methods are able to detect the majority of clear pseudogenes, manual annotation is necessary to grasp the subtleties of the more obscure types of pseudogenes.

Non-Coding RNAs

In humans, non-coding RNAs (ncRNA) may equal or be up to three times higher than the number of protein coding transcripts, even more if introns are taken into account (Mattick 2001, 2003). When predicting gene numbers and annotating genes, this class has been consistently overlooked because focus is placed on protein-coding genes (Wright *et al.* 2001). The reasons for this are the limitations of the current three methods used for gene estimation and annotation: sequencing of mRNAs through cDNAs and ESTs (Ewing and Green 2000, Liang *et al.* 2000, Kawai *et al.* 2001, Okazaki *et al.* 2002), computational prediction (Lander *et al.* 2001, Venter *et al.* 2001) and comparative genome analysis identifying conserved

ORFs (Crollius *et al.* 2000). These methods are biased towards polyadenylated protein coding transcripts and do not work well for ncRNA genes, which produce transcripts that function as structural, catalytic or regulatory RNAs and are often small and not polyadenylated. Although many genomes have been sequenced and annotated, both the number and diversity of ncRNA genes remain unknown quantities. Microarray experiments identify many more expressed sequences on the genome than currently annotated (Shoemaker *et al.* 2001, Kapranov *et al.* 2002, Saha *et al.* 2002, Bertone *et al.* 2004, Kampa *et al.* 2004, Schadt *et al.* 2004), a substantial fraction of which could represent ncRNAs.

Non-coding RNAs come in a variety of forms (Eddy 2001, Storz 2002, Szymanski and Barciszewski 2002) and therefore require different gene-finding approaches. Programs like QRNA (Rivas and Eddy 2001), DDBRNA (di Bernardo *et al.* 2003), MSARI (Coventry *et al.* 2004) and RNAZ (Washietl *et al.* 2005) use pairwise or multiple genome comparisons to predict regions potentially coding for structurally conserved mRNAs (Rivas *et al.* 2001). Other programs are specifically geared towards finding one class of ncRNAs, for example tRNAs (Fichant and Burks 1991, Pavesi *et al.* 1994). Different types of ncRNAs, such as tRNAs, rRNAs, small nucleolar RNAs and small nuclear RNAs perform many different, specialized, functions, for example in the translation machinery (tRNA, rRNA), the splicing machinery and gene regulation (Eddy 2001). One class of ncRNAs that appear to have a large impact on the function of an organism is antisense RNAs. The XIST antisense ncRNA is thought to be involved in silencing of X-linked genes by the modification of chromatin (Boumil and Lee 2001, Cohen and Lee 2002). Other antisense RNA examples found within imprinted clusters include the IPW (imprinted in Prader-Willi syndrome) and H19 (imprinted maternally expressed untranslated mRNA) transcripts (Rougeulle and Heard 2002). Detailed re-annotation of the finished chromosomes 22 has revealed 16 potential cis-antisense RNAs, *i.e.* RNAs that overlap coding genes on the opposite genomic strand (Collins *et al.* 2003). MicroRNAs (miRNAs), small non-coding RNAs processed from larger transcripts, are trans-antisense RNAs regulating and binding to mRNAs (Bartel and Chen 2004, He and Hannon 2004).

As both experimental and computational biologists devise new strategies to specifically detect eukaryotic ncRNAs, annotation of these structures will greatly improve. One can, for example, find many non-coding RNA genes in the Rfam database and use the models contained therein to scan genomes for potential ncRNA genes (Griffiths-Jones *et al.* 2003). The growing importance of ncRNAs shows that to fully understand regulatory genetics we have to look beyond the proteome when analysing the genome.

Complex Loci

Genomic loci with complex gene arrangements and clusters of conserved duplicated genes are challenging to automated annotation systems. Problems often occur with correct naming of genes when two or more virtually identical genes are located in close vicinity, as for example in the HIST1 histone gene cluster in human or mouse. This cluster contains multiple copies of all five histone genes, with every copy of a single type giving rise to an identical translation. Thus it is near impossible to correctly name the individual loci and detailed manual annotation is needed for correct identification. At the time of writing the Ensembl-assigned gene names are markedly different (with many names occurring more than once) from the corresponding Vega-assigned names. Additionally, the proper classification of some of these genes as pseudogenes tends to fall outside the capabilities of the automated approach (see more on this in the "Pseudogenes" section).

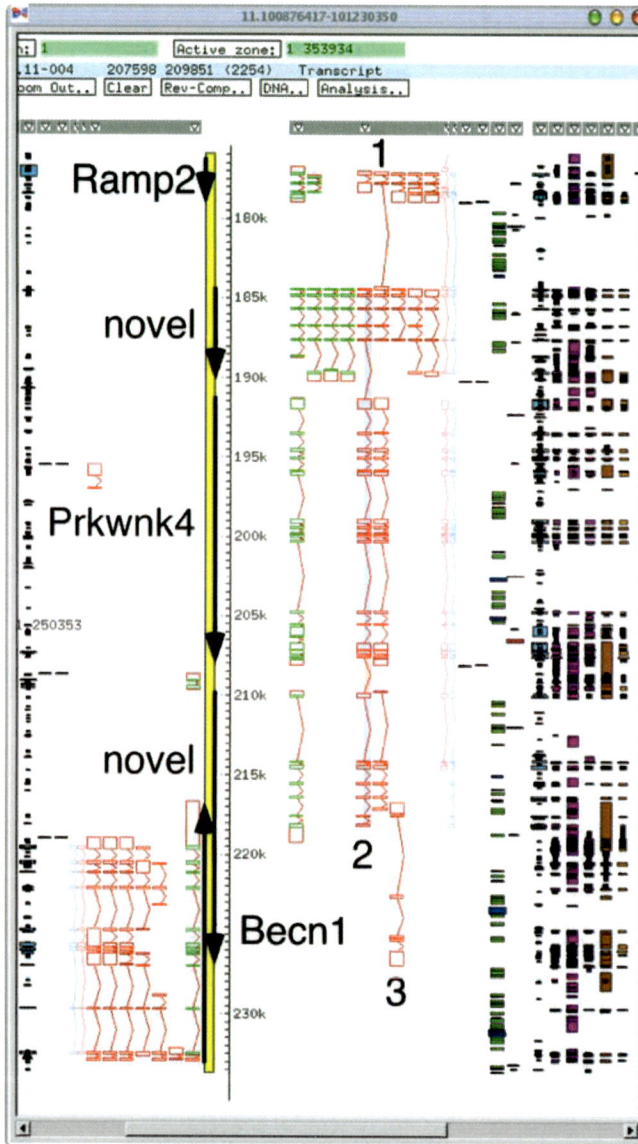

Figure 2. A complex locus on mouse chromosome 11, seen in the Otterlace annotation interface. Transcript 1 joins the Ramp2 and the downstream novel gene. Transcript 2 uses exons from the novel gene, gene Prkwnk2 and another novel gene. This transcript is based on a rat mRNA and uses unique canonical splice donor sites at the gene junctions. Also note Transcript 3, extending from the novel gene and overlapping with the Becn1 gene on the opposite strand. None of the three transcripts mentioned have a believable CDS (which would be shown as green outline exons), though Transcript 2 has a full-length CDS incorporating all three genes in its native species (rat).

Apart from gene clusters, complex gene arrangements usually require a human eye to untangle the intricate relationships between various transcripts and loci. Two examples involving what can variously be referred to as gene-fusion, poly-cistronic mRNAs or transcriptional read-through can serve as an illustration. Genes Tnfsf12 (OTTMUSG00000005999) and Tnfsf13 (OTTMUSG00000006001) are located next to each other on mouse chromosome 11 (and on the syntenic human 17). They appear to be independent individual genes that are in no way similar (*i.e.* did not arise from a duplication), but one mRNA exists that joins up exons of both genes and gives rise to a translation incorporating most of both proteins (only the last exon of the upstream Tnfsf12 gene and the first of the downstream Tnfsf13 gene are skipped), suggesting that maybe the two genes are splice variants of a single gene, even though the two translations have no overlap and the genes have their own 5' and 3' features. Another example of transcripts linking two or more genes, but in this case with the much more common lack of a CDS across the loci, is the group of Ramp2, Prkwnk4 and other loci on mouse chromosome 11 (figure 2).

Here a transcript, based on EST BF577527, is processed from exons of the upstream Ramp2 gene and gene OTTMUSG00000002759 following it, while a second mRNA (actually from the rat, AY321328) is processed from most exons belonging to the latter gene and genes Prkwnk4 (OTTMUSG00000002761) and OTTMUSG00000002763. This second transcript uses unique splice acceptor sites (*i.e.* not used by the gene-specific transcripts) at the gene junctions. Neither of the transcripts has a CDS joining the various individual gene translations, though interestingly the rat-derived transcript does have a full CDS in the rat. The complexity of this locus set is increased by the presence of transcript AK014884, a variant of the most downstream of the four loci that overlaps with the Becn1 (OTTMUSG00000002791) gene on the opposite strand.

There are naturally many more examples of these types of complex gene structures and arrangements that understandably tend to confound, or be overlooked by, automated systems. Owing to the manual annotation that serves as the source for Vega data, Vega will accurately show the full complexity of these genomic regions.

Comparative Genome Analysis and Annotation

Homologous genes are derived from a common ancestral gene and their level of sequence similarity indicates the rate of divergence from their common ancestor. Two types of homologs can be defined: orthologs, which are derived from the same gene in the last common ancestral species and thus have similar functions, and paralogs, which are produced by duplication of a chromosomal segment and typically have different (but often overlapping) functions. It is now well established that comparative analysis of genomic sequence from species at different evolutionary distances is a powerful method for identifying coding sequences, conserved non-coding sequences with regulatory functions and for highlighting unique species-specific sequences (Crollius *et al.* 2000, Wasserman *et al.* 2000, Touchman *et al.* 2001, Waterston *et al.* 2002, Frazer *et al.* 2003, Mallon *et al.* 2004). Comparison of the human and the mouse genome sequences can be used to refine and expand our knowledge of the contents of both genomes. From initial comparative analysis it was found that approximately 80% of mouse genes have a single identifiable ortholog in the human genome and less than 1% are without a detectable human homolog.

As mentioned earlier, Vega can be used to browse analysis and annotation of any size contiguous or non-contiguous genomic sequence. An example is the various MHC (Major Histocompatibility Complex) regions that have been sequenced and annotated. For the human MHC region, the previous mosaic of genomic sequences in the original chromosome 6 assembly has been replaced with that of a single haplotype (Horton *et al.* 2004). The human MHC contains many immune related genes, including highly polymorphic genes for MHC class I and class II molecules that present antigens to T lymphocytes. Genetic determinants for many autoimmune diseases (for example type 1 diabetes and multiple sclerosis) and some infectious diseases have been linked to the MHC, but the precise determinants have proven difficult to identify. Detailed annotation of the region, and especially hypervariable sub-regions such as RCCX and DRB, is a fundamental first step towards finding the genes involved in these diseases. In addition to the annotation of the reference haplotype chosen as part of the whole chromosome 6 sequence, the MHC regions of seven other distinct human haplotypes are being sequenced and annotated, all of which will be available through Vega. Compara view allows users to compare the different haplotypes, find the variations between them and their ancestral relationships, which may accelerate identification of disease associated MHC genes. Likewise, MHC regions from different organisms can be drawn into this comparison, because the equivalent region has been sequenced in several different organisms, such as cat, dog, pig and chicken, all of which will be made available through Vega.

Other special regions one can find in Vega are regions of the mouse genome that have been annotated specifically as part of research into regions linked to diseases or developmental anomalies. One example is the Del36H region on mouse chromosome 13 (Mallon *et al.* 2004). Several disease loci are located in the deletion (Davies *et al.* 1999) and heterozygous loss of the region causes phenotypes that may model human genetic diseases. A remarkable number of gene clusters map to the Del36H region: less than 13 Mb of genomic sequence contains vomeronasal receptor, prolactin, serpin (serine protease inhibitor), butyrophilin and histone gene clusters, as well as several smaller ones such as Slc (solute carrier) and Fox (forkhead box). In-depth annotation allowed detailed comparison of the human and mouse sequences, revealing the precise differences in gene arrangements between the two (Mallon *et al.* 2004). In mouse most of the large clusters have expanded greatly (between 7-fold and 26-fold), except the butyrophilin cluster which has expanded from two members in mouse to seven in human (figure 3). The histone cluster is not much different in size and organization between the two species, but several family members unique (positionally, not protein sequence-wise) to each species were identified, as well as members degenerated into pseudogenes in only one of the species.

Another example of dedicated regions is the set of IDD candidate regions on the mouse genome. The Non-Obese Diabetic (NOD) mouse strain is a strain susceptible to develop Insulin Dependent Diabetes (IDD) (Hill *et al.* 2000). Several candidate regions for genes involved in IDD susceptibility have been identified on chromosomes 1 and 3, and the annotation for these regions will be available for both the NOD strain and the reference C57BL/6 strain. Thus one can compare the sequence and genes between the susceptible and the non-susceptible strains (for example comparing SNPs (Wicker *et al.* 2004)) in order to try and identify the susceptibility genes.

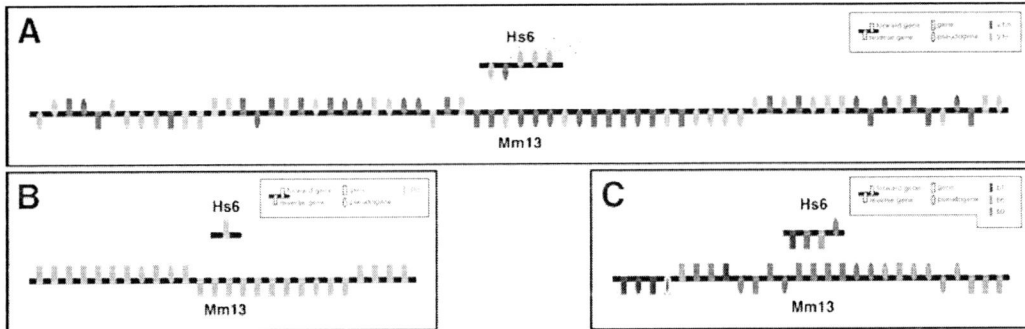

Figure 3. Detailed annotation of three clusters on mouse chromosome 13 that appear to have expanded in mouse compared to human. Coloured blocks depict genes transcribed in the forward (up) or reverse (down) direction. Ovals represent pseudogenes. Figure is not to scale and any non-cluster genes that may be interspersed are not shown. A: the vomeronasal receptor cluster (V1rh and V1ri) in mouse consists of 67 members including 34 pseudogenes, while the equivalent human cluster on chromosome 6 has only five pseudogenes, two of which cannot be specifically assigned to family h or i (orange ovals). B: the single prolactin (PRL) gene in human has expanded to a 26 member cluster in mouse. C: a group of serine proteinase inhibitor genes consisting of one member each of SERPINB1, SERPINB6 and SERPINB9 in human has undergone multiple local duplications in mouse while maintaining to a large extend the overall gene order.

Data gained from comparing genomic sequences from two species can be used to improve gene-structure predictions. TWINSCAN (Korf *et al.* 2001), SGP-1 (syntenic gene prediction) (Wiehe *et al.* 2001) and SLAM (Alexandersson *et al.* 2003) have been developed for this purpose, using a combination of pairwise similarity and *ab initio* gene prediction. Manually annotated sequence can be used as a reference or test set to determine which combination of comparative data is most useful. Groups involved in the manual annotation of human chromosomes 14 and 20 and re-annotation of 22, have examined various comparative strategies for improving or confirming gene predictions. They found that a combination of mouse whole genome shotgun reads and *Tetraodon* genomic evolutionary conserved regions detected almost all annotated protein coding exons on chromosome 20 (Deloukas *et al.* 2001) and that only eleven putative new genes were predicted on chromosome 14 (Heilig *et al.* 2003). As discussed in the chromosome 22 re-annotation paper (Collins *et al.* 2003) and elsewhere in this chapter, Vega gene structures are usually constructed based on cDNA derived homology data and exclude exons not supported by transcript or protein homology. Nevertheless, cross species analysis and gene prediction programs are useful for identifying and confirming expressed regions of the genome and highlighting cases where annotation is incomplete in either species.

Accurate and detailed comparison of various genomes and their gene content, especially between mouse and human, helps further our insight into the function of genes and our ability to determine which genes of biomedical interest in human to experimentally manipulate in model organisms.

Experimental Validation

Manual annotation is an important starting point from where to complete the transcriptome. A proportion of annotated genes is not completely, or not at all, represented by expressed sequence, despite the various large scale cDNA sequencing projects underway. Instead of starting with the transcriptome and mapping that to the genome, Collins *et al.* (2004) started with the genome (annotated exons on human chromosome 22) to add to the transcriptome or ORFeome. On a smaller scale, the Encode (encyclopedia of DNA elements) pilot project (Feingold *et al.* 2004) starts with manual annotation of selected regions (some disease associated, some randomly selected) representing in total around 1% of the human genome and subsequently tries to isolate a complete set of corresponding transcripts and identify all relevant functional elements with the help of the syntenic regions from several species. The Encode project will result in improved annotation methods by enhancing our understanding of gene structures and the results will be presented through Vega.

Future Direction of Gene Annotation

Whole genome annotation is a continual process and we are still in the primary phase. The CCDS project will help define a standardised reference set of human protein-coding genes that will become the scaffold for additional functional annotation to build on. The sequencing and annotation of different haplotypes over selected regions of the genome will help understand genes involved in complex diseases such as diabetes and schizophrenia. Incorporation of results from high-throughput experiments such as microarray expression analysis, yeast two-hybrid, long range serial analysis of gene expression (SAGE) or green fluorescent protein tagging of proteins will add further functional information to enhance the gene annotation. Over the next few years the aim will be to integrate all types of computational, curated and experimental data in a single repository enabling the biologist to look at a gene sequence and determine its biological role within the cell.

References

Alexandersson, M., Cawley, S. and Pachter, L. (2003) SLAM: cross-species gene finding and alignment with a generalized pair hidden Markov model. *Genome Res.*, 13, 496-502.

Anderson, C.L., Zundel, M.A. and Werner, R. (2005) Variable promoter usage and alternative splicing in five mouse connexin genes. *Genomics*, 85, 238-244.

Ashurst, J.L. and Collins, J.E. (2003) Gene annotation: prediction and testing. *Annu Rev Genomics Hum Genet.*, 4, 69-88.

Bartel, D.P. and Chen, C.Z. (2004) Micromanagers of gene expression: the potentially widespread influence of metazoan microRNAs. *Nat Rev Genet.* 5, 396-400.

Batzoglou, S., Pachter, L., Mesirov, J.P., Berger, B. and Lander, E.S. (2000) Human and mouse gene structure: comparative analysis and application to exon prediction. *Genome Res.*, 10, 950-958.

Beaudoing, E., Freier, S., Wyatt, J.R., Claverie, J.M. and Gautheret, D. (2000) Patterns of variant polyadenylation signal usage in human genes. *Genome Res.*, 10, 1001-1010.

di Bernardo, D., Down, T. and Hubbard, T. (2003) ddbRNA: detection of conserved secondary structures in multiple alignments. *Bioinformatics*, 19, 1606-1611.

Bertone, P., Stolc, V., Royce, T. E., Rozowsky, J. S., Urban, A. E., Zhu, X., Rinn, J. L., Tongprasit, W., Samanta, M., Weissman, S. *et al.* (2004) Global identification of human transcribed sequences with genome tiling arrays. *Science*, 306, 2242-2246.

Birney, E. and Durbin, R. (2000) Using GeneWise in the Drosophila annotation experiment. *Genome Res.*, 10, 547-548.

Boumil, R.M. and Lee, J.T. (2001) Forty years of decoding the silence in X-chromosome inactivation. *Hum Mol Genet.*, 10, 2225-2232.

Brett, D., Pospisil, H., Valcarcel, J., Reich, J. and Bork, P. (2002) Alternative splicing and genome complexity. *Nat Genet.*, 30, 29-30.

Burge, C. and Karlin, S. (1997) Prediction of complete gene structures in human genomic DNA. *J Mol Biol.*, 268, 78-94.

Cohen, D.E. and Lee, J.T. (2002) X-chromosome inactivation and the search for chromosome-wide silencers. *Curr Opin Genet Dev.*, 12, 219-224.

Collins, J.E., Goward, M.E., Cole, C.G., Smink, L.J., Huckle, E.J., Knowles, S., Bye, J.M., Beare, D.M. and Dunham, I. (2003) Reevaluating human gene annotation: a second-generation analysis of chromosome 22. *Genome Res.*, 13, 27-36.

Collins, J.E., Wright, C.L., Edwards, C.A., Davis, M.P., Grinham, J.A., Cole, C.G., Goward, M.E., Aguado, B., Mallya, M., Mokrab, Y. *et al.* (2004) A genome annotation-driven approach to cloning the human ORFeome. *Genome Biol.*, 5, R84.1-R84.11.

Cooke, J., Nowak, M.A., Boerlijst, M. and Maynard-Smith, J. (1997) Evolutionary origins and maintenance of redundant gene expression during metazoan development. *Trends Genet.*, 13, 360-364.

Coventry, A., Kleitman, D. J. and Berger, B. (2004) MSARI: Multiple sequence alignments for statistical detection of RNA secondary structure. *Proc Natl Acad Sci U S A.*, 101, 12102-12107.

Crollius, H., Jaillon, O., Bernot, A., Dasilva, C., Bouneau, L., Fischer, C., Fizames, C., Winckcr, P., Brottier, P., Quetier, F. *et al.* (2000) Estimate of human gene number provided by genome-wide analysis using Tetraodon nigroviridis DNA sequence. *Nat Genet.*, 25, 235-238.

Daniels, G.R. and Deininger, P.L. (1985) Repeat sequence families derived from mammalian tRNA genes. *Nature*, 317, 819–822.

Davies, A.F., Mirza, G., Sekhon, G., Turnpenny, P., Leroy, F., Speleman, F., Law, C., van Regemorter, N., Vamos, E., Flinter, F. *et al.* (1999) Delineation of two distinct 6p deletion syndromes. *Hum Genet.*, 104, 64-72.

Deloukas, P., Matthews, L.H., Ashurst, J., Burton, J., Gilbert, J.G., Jones, M., Stavrides, G., Almeida, J.P., Babbage, A.K., Bagguley, C.L. *et al.* (2001) The DNA sequence and comparative analysis of human chromosome 20. *Nature*, 414, 865-871.

Dunham, I., Shimizu, N., Roe, B.A., Chissoe, S., Hunt, A.R., Collins, J.E., Bruskiewich, R., Beare, D.M., Clamp, M., Smink, L.J. *et al.* (1999) The DNA sequence of human chromosome 22. *Nature*, 402, 489-495.

Eddy, S.R. (2001) Non-coding RNA genes and the modern RNA world. *Nat Rev Genet.*, 2, 919-929.

Edwalds-Gilbert, G., Veraldi, K.L. and Milcarek, C. (1997) Alternative poly(A) site selection in complex transcription units: means to an end? *Nucl Acids Res.*, 25, 2547-2561.

Ewing, B. and Green, P. (2000) Analysis of expressed sequence tags indicates 35,000 human genes. *Nat Genet.*, 25, 232-234.

Feingold, E.A., Good, .P.J., Guyer, M.S., Kamholz, S., Liefer, L., Wetterstrand, K., Collins, F.S. *et al.* (2004) The ENCODE (ENCyclopedia Of DNA Elements) project. *Science*, 306, 636-640.

Fichant, G.A. and Burks, C. (1991) Identifying potential tRNA genes in genomic DNA sequences. *J Mol Biol.*, 220, 659-671.

Frazer, K.A., Elnitski, L., Church, D.M., Dubchak, I. and Hardison, R.C. (2003) Cross-species sequence comparisons: a review of methods and available resources. *Genome Res.*, 13, 1-12.

Gaasterland, T. and Oprea, M. (2001) Whole-genome analysis: annotations and updates. *Curr Opin Struct Biol.*, 11, 377-381.

Ganfornina, M.D. and Sanchez, D. (1999) Generation of evolutionary novelty by functional shift. *Bioessays*, 21, 432-439.

Graveley, B. (2001) Alternative splicing: increasing diversity in the proteomic world. *Trends Genet.*, 17, 100-107.

Griffiths-Jones, S., Bateman, A., Marshall, M., Khanna, A. and Eddy, S.R. (2003) Rfam: an RNA family database. *Nucl Acids Res.*, 31, 439-441.

Harrison, P.M., Hegyi, H., Balasubramanian, S., Luscombe, N.M., Bertone, P., Echols, N., Johnson, T. and Gerstein, M. (2002) Molecular fossils in the human genome: Identification and analysis of the pseudogenes in chromosomes 21 and 22. *Genome Res.*, 12, 272-280.

Hattori, M., Fujiyama, A., Taylor, T.D., Watanabe, H., Yada, T., Park, H.S., Toyoda, A., Ishii, K., Totoki, Y., Choi, D.K. *et al.* (2000) The DNA sequence of human chromosome 21. *Nature*, 405, 311-319.

He, L. and Hannon, G.J. (2004) MicroRNAs: small RNAs with a big role in gene regulation. *Nat Rev Genet.*, 5, 522-531.

Heilig, R., Eckenberg, R., Petit, J.L., Fonknechten, N., Da Silva, C., Cattolico, L., Levy, M., Barbe, V., De Berardinis, V., Ureta-Vidal, A. *et al.* (2003) The DNA sequence and analysis of human chromosome 14. *Nature*, 421, 601-607.

Hide, W.A., Babenko, V.N., van Heusden, P.A., Seoighe, C. and Kelso, J.F. (2001) The contribution of exon-skipping events on chromosome 22 to protein coding diversity. *Genome Res.*, 11, 1848-1853.

Hill, N.J., Lyons, P.A., Armitage, N., Todd, J.A., Wicker, L.S., Peterson, L.B. (2000) NOD Idd5 locus controls insulitis and diabetes and overlaps the orthologous CTLA4/IDDM12 and NRAMP1 loci in humans. *Diabetes*, 49, 1744-1747.

Hirotsune, S., Yoshida, N., Chen, A., Garrett, L., Sugiyama, F., Takahashi, S., Yagami, K.-I., Wynshaw-Boris, A. and Yoshik, A. (2003) An expressed pseudogene regulates the messenger-RNA stability of its homologous coding gene. *Nature*, 423, 91-96.

Horton, R., Wilming, L., Rand, V., Lovering, R.C., Bruford, E.A., Khodiyar, V.K., Lush, M.J., Povey, S., Talbot Jr., C.C., Wright, M.W. *et al.* (2004) Gene map of the extended human MHC. *Nat Rev Genet.*, 5, 889-899.

Hubbard, T., Barker, D., Birney, E., Cameron, G., Chen, Y., Clark, L., Cox, T., Cuff, J., Curwen, V., Down, T. *et al.* (2002) The Ensembl genome database project. *Nucl Acids Res.*, 30, 38-41.

Kampa, D., Cheng, J., Kapranov, P., Yamanaka, M., Brubaker, S., Cawley, S., Drenkow, J., Piccolboni, A., Bekiranov, S., Helt, G. *et al.* (2004) Novel RNAs identified from an in-depth analysis of the transcriptome of human chromosomes 21 and 22. *Genome Res.*, 14, 331-342.

Kan, Z., States, D., Gish, W., Rouchka, E.C., Gish, W.R., States, D.J., Rouchka, E. and Glasscock, J. (2002) Selecting for Functional Alternative Splices in ESTs Gene structure prediction and alternative splicing analysis using genomically aligned ESTs UTR reconstruction and analysis using genomically aligned EST sequences. *Genome Res.*, 12, 1837-1845.

Kapranov, P., Cawley, S.E., Drenkow, J., Bekiranov, S., Strausberg, R.L., Fodor, S.P. and Gingeras, T.R. (2002) Large-scale transcriptional activity in chromosomes 21 and 22. *Science*, 296, 916-919.

Kawai, J., Shinagawa, A., Shibata, K., Yoshino, M., Itoh, M., Ishii, Y., Arakawa, T., Hara, A., Fukunishi, Y., Konno, H. *et al.* (2001) Functional annotation of a full-length mouse cDNA collection. *Nature*, 409, 685-690.

Kenmochi, N., Yoshihama, M., Higa, S. and Tanaka, T. (2000) The human ribosomal protein L6 gene in a critical region for Noonan syndrome. *J Hum Genet.*, 45, 290-293.

Korf, I., Flicek, P., Duan, D. and Brent, M.R. (2001) Integrating genomic homology into gene structure prediction. *Bioinformatics*, 17 Suppl 1, S140-148.

Lander, E.S., Linton, L.M., Birren, B., Nusbaum, C., Zody, M.C., Baldwin, J., Devon, K., Dewar, K., Doyle, M., FitzHugh, W. *et al.* (2001) Initial sequencing and analysis of the human genome. *Nature*, 409, 860-921.

Lewis, B.P., Green, R.E. and Brenner, S.E. (2002) Evidence for the widespread coupling of alternative splicing and nonsense-mediated mRNA decay in humans. *Proc Natl Acad Sci U S A.*, 100, 189-192.

Liang, F., Holt, I., Pertea, G., Karamycheva, S., Salzberg, S.L. and Quackenbush, J. (2000) Gene index analysis of the human genome estimates approximately 120,000 genes. *Nat Genet.*, 25, 239-240.

Lopez, A.J. (1998) Alternative splicing of pre-mRNA: developmental consequences and mechanisms of regulation. *Annu Rev Genet.*, 32, 279-305.

Mallon, A.-M., Wilming, L., Weekes, J., Gilbert, J.G.R., Ashurst, J., Peyrefitte, S., Matthews, L., Cadman, M., McKeone, R., Sellick, C.A. *et al.* (2004) Organization and Evolution of a Gene-Rich Region of the Mouse Genome: A 12.7-Mb Region Deleted in the Del(13)*Svea*36H Mouse. *Genome Res*, 14, 1888-1901

Mattick, J.S. (2001) Non-coding RNAs: the architects of eukaryotic complexity. *EMBO Rep.*, 2, 986-991.

Mattick, J.S. (2003) Challenging the dogma: the hidden layer of non-protein-coding RNAs in complex organisms. *BioEssays*, 25, 930-939.

McCarrey, J.R. and Thomas, K. (1987) Human testis-specific PGK gene lacks introns and possesses characteristics of a processed gene. *Nature*, 326, 501-505.

Modrek, B. and Lee, C. (2002) A genomic view of alternative splicing. *Nat Genet.*, 30, 13-19.

Okazaki, Y., Furuno, M., Kasukawa, T., Adachi, J., Bono, H., Kondo, S., Nikaido, I., Osato, N., Saito, R., Suzuki, H. *et al.* (2002) Analysis of the mouse transcriptome based on functional annotation of 60,770 full-length cDNAs. *Nature*, 420, 563-573.

Pavesi, A., Conterio, F., Bolchi, A., Dieci, G. and Ottonello, S. (1994) Identification of new eukaryotic tRNA genes in genomic DNA databases by a multistep weight matrix analysis of trnascriptional control regions. *Nucl Acids Res.*, 22, 1247-1256.

Pesole, G., Liuni, S., Grillo, G., Licciulli, F., Mignone, F., Gissi, C. and Saccone, C. (2002) UTRdb and UTRsite: specialized databases of sequences and functional elements of 5' and 3' untranslated regions of eukaryotic mRNAs. Update 2002. *Nucl Acids Res.*, 30, 335-340.

Pruitt, K.D. and Maglott, D.R. (2001) RefSeq and LocusLink: NCBI gene-centered resources. *Nucleic Acids Res.*, 29, 137-140.

Reese, M.G., Hartzell, G., Harris, N.L., Ohler, U., Abril, J.F. and Lewis, S.E. (2000) Genome annotation assessment in Drosophila melanogaster. *Genome Res.*, 10, 483-501.

Reese, M.G., Kulp, D., Tammana, H. and Haussler, D. (2000) Genie - Gene finding in Drosophila melanogaster. *Genome Res.*, 10, 529-538.

Rivas, E. and Eddy, S.R. (2001) Noncoding RNA gene detection using comparative sequence analysis. *BMC Bioinformatics*, 2, 8.

Rivas, E., Klein, R.J., Jones, T.A. and Eddy, S.R. (2001) Computational identification of noncoding RNAs in E. coli by comparative genomics. *Curr Biol.*, 11, 1369-1373.

Rogic, S., Mackworth, A.K. and Ouellette, F.B. (2001) Evaluation of gene-finding programs on mammalian sequences. *Genome Res.*, 11, 817-832.

Rougeulle, C. and Heard, E. (2002) Antisense RNA in imprinting: spreading silence through Air. *Trends Genet.*, 18, 434-437.

Saha, S, Sparks, A.B., Rago, C., Akmaev, V., Wang, C.J., Vogelstein, B., Kinzler, K.W. and Velculescu, V.E. (2002) Using the transcriptome to annotate the genome. *Nat Biotechnol.*, 20, 508-512.

Schadt, E.E., Edwards, S.W., GuhaThakurta, D., Holder, D., Ying, L., Svetnik, V., Leonardson, A., Hart, K.W., Russell, A., Li, G. *et al.* (2004) A comprehensive transcript index of the human genome generated using microarrays and computational approaches. *Genome Biol.*, 5, R73.1-R73.17.

Schmucker, D., Clemens, J.C., Shu, H., Worby, C.A., Xiao, J., Muda, M., Dixon, J.E. and Zipursky, S.L. (2000) Drosophila Dscam is an axon guidance receptor exhibiting extraordinary molecular diversity. *Cell*, 101, 671-684.

Searle, S.M., Gilbert, J., Iyer, V. and Clamp, M. (2004) The otter annotation system. *Genome Res.*, 14, 963-970.

Shoemaker, D.D., Schadt, E.E., Armour, C.D., He, Y.D., Garrett-Engele, P., McDonagh, P.D., Loerch, P.M., Leonardson, A., Lum, P.Y., Cavet, G. *et al.* (2001) Experimental annotation of the human genome using microarray technology. *Nature*, 409, 922-927.

Smith, C.W. and Valcarcel, J. (2000) Alternative pre-mRNA splicing: the logic of combinatorial control. *Trends Biochem Sci.*, 25, 381-388.

Solovyev, V.V., Salamov, A.A. and Lawrence, C.B. (1995) Identification of human gene structure using linear discriminant functions and dynamic programming. *Proc Int Conf Intell Syst Mol Biol.*, 3, 367-375.

Stamm, S., Ben-Ari, S., Rafalska, I., Tang, Y., Zhang, Z., Toiber, D., Thanaraj, T.A. and Soreq, H. (2005) Function of alternative splicing. *Gene*, 344, 1-20.

Storz, G. (2002) An expanding universe of noncoding RNAs. *Science*, 296, 1260-1263.

Szymanski, M. and Barciszewski, J. (2002) Beyond the proteome: non-coding regulatory RNAs. *Genome Biol.*, 3, reviews0005.1-reviews0005.8.

Tabaska, J.E. and Zhang, M.Q. (1999) Detection of polyadenylation signals in human DNA sequences. *Gene*, 231, 77-86.

Taneri, B., Snyder, B., Novoradovsky, A. and Gaasterland, T. (2004) Alternative splicing of mouse transcription factors affects their DNA-binding domain architecture and is tissue specific. *Genome Biol.*, 5, R75.1-R75.9.

Touchman, J.W. Dehejia, A., Chiba-Falek, O., Cabin, D.E., Schwartz, J.R., Orrison, B.M., Polymeropoulos, M.H. and Nussbaum, R.L. (2001) Human and mouse □-synuclein genes: comparative genomic sequence analysis and identification of a novel gene regulatory element. *Genome Res.*,11, 78-86.

Vanin, E.F. (1985) Processed pseudogenes: characteristics and evolution. *Annu Rev Genet.*, 19, 253-272.

Venter, J.C., Adams, M.D., Myers, E.W., Li, P.W., Mural, R.J., Sutton, G.G., Smith, H.O., Yandell, M., Evans, C.A., Holt, R.A. *et al.* (2001) The sequence of the human genome. *Science*, 291, 1304-1351.

Wain, H.M., Lush, M.J., Ducluzeau, F., Khodiyar, V.K. and Povey, S. (2004) Genew: the Human Gene Nomenclature Database, 2004 updates. *Nucl Acids Res.*, 32 Database issue, D255-257.

Washietl, S., Hofacker, I.L. and Stadler, P.F. (2005) Fast and reliable prediction of noncoding RNAs. *Proc Natl Acad Sci U S A.*, 102, 2454-2459.

Wasserman, W.W., Palumbo, M., Thompson, W., Fickett, J.W. and Lawrence, C.E. (2002) Human-mouse genome comparisons to locate regulatory sites. *Nat Genet.*, 26, 225-228.

Waterston, R.H., Lindblad-Toh, K., Birney, E., Rogers, J., Abril, J.F., Agarwal, P., Agarwala, R., Ainscough, R., Alexandersson, M., An, P. *et al.* (2002) Initial sequencing and comparative analysis of the mouse genome. *Nature*, 420, 520-562.

Weiner, A.M., Deininger, P.L. and Efstratiadis, A. (1986) Nonviral retroposons: genes, pseudogenes, and transposable elements generated by the reverse flow of genetic information. *Annu Rev Biochem.*, 55, 631–661.

Wicker, L.S., Chamberlain, G., Hunter, K., Rainbow, D., Howlett, S., Tiffen, P., Clark, J., Gonzalez-Munoz, A., Cumiskey, A.M., Rosa R.L. *et al.* (2004) Fine Mapping, Gene Content, Comparative Sequencing, and Expression Analyses Support *Ctla4* and *Nramp1* as Candidates for *Idd5.1* and *Idd5.2* in the Nonobese Diabetic Mouse. *J Immunol.*, 173, 164-173.

Wiehe, T., Gebauer-Jung, S., Mitchell-Olds, T. and Guigo, R. (2001) SGP-1: prediction and validation of homologous genes based on sequence alignments. *Genome Res.*, 11, 1574-1583.

Wright, F.A., Lemon, W.J., Zhao, W.D., Sears, R., Zhuo, D., Wang, J.P., Yang, H.Y., Baer, T., Stredney, D., Spitzner, J. *et al.* (2001) A draft annotation and overview of the human genome. *Genome Biol.*, 2, research0025.1- research0025.18.

Yamakawa, K., Huot, Y.K., Haendelt, M.A., Hubert, R., Chen, X.N., Lyons, G.E. and Korenberg, J.R. (1998) DSCAM: a novel member of the immunoglobulin superfamily maps in a Down syndrome region and is involved in the development of the nervous system. *Hum Mol Genet.*, 7, 227-237.

Yeh, R.F., Lim, L.P. and Burge, C.B. (2001) Computational inference of homologous gene structures in the human genome. *Genome Res.*, 11, 803-816.

Yeo, G., Holste, D., Kreiman, G. and Burge, C.B. (2004) Variation in alternative splicing across human tissues. *Genome Biol.*, 5, R74.1-R74.15.

Yeo, G.W., Van Nostrand, E., Holste, D., Poggio, T. and Burge, C.B. (2005) Identification and analysis of alternative splicing events conserved in human and mouse. *Proc Natl Acad Sci U S A.*, 102, 2850-2855.

Zhang, M.Q. (1997) Identification of protein coding regions in the human genome by quadratic discriminant analysis. *Proc Natl Acad Sci U S A.*, 94, 565-568.

In: In Silico Genomics and Proteomics
Editors: N. Mulder and R. Apweiler, pp. 147-156

ISBN 1-59454-995-8
© 2006 Nova Science Publishers, Inc.

Chapter 11

The PEDANT Genome Database

Dmitrij Frishman[1,3]*, Klaus Heumann[2] and Hans-Werner Mewes[1,3]
[1]Institute for Bioinformatics, GSF - National Research Center for Health and
Environment, Ingolstädter Landstraße 1, 85764 Neuherberg, Germany
[2]Biomax Informatics AG, Lochhamer Straße 9, 82152 Martinsried, Germany
[3]Department of Genome-oriented Bioinformatics, Wissenschaftszentrum Weihenstephan,
Technische Universität München, 85350 Freising, Germany

Abstract

The PEDANT Genome Database (http://pedant.gsf.de) contains pre-computed information resulting from bioinformatics analyses of publicly available genomes. Its main mission is to provide robust automatic annotation of the vast majority of amino acid sequences which have not been subjected to in-depth manual curation by human experts in high-quality protein sequence databases. By design PEDANT annotation is genome-oriented, making it possible to explore genomic context of gene products and evaluate functional and structural content of genomes using a category-based query mechanism. At present the PEDANT database contains exhaustive annotation of over 1 240 000 proteins from 297 eubacterial, 23 archeal, and 51 eukaryotic genomes.

Keywords: genome analysis, sequence annotation, data integration, structural genomics

Introduction

One of the most valuable information sources in modern biology is constituted by manually curated molecular biological databanks. Those include the generic protein sequence database Uniprot [Bairoch *et al.*, 2005], genome databases of selected model organisms [Keseler *et al.*, 2005; Guldener *et al.*, 2005], collections of sequence domains [Bateman *et al.*,

---
* Corresponding author. E-mail: d.frishman@wzw.tum.de; fax: +49-8161-712186

2004], as well as a number of resources specializing on selected aspects of protein function, such as protein-protein interactions (e.g., [Pagel *et al.*, 2004]). These and many other databanks serve as important reference points to guide and validate high-throughput experiments and computer-based function predictions.

Unfortunately, it is becoming increasingly clear that most of the protein sequences will never be annotated by human experts. The gap between the number of entries in Uniprot/Swissprot (roughly 150.000 at the time of writing) and the total number of known sequences (3 mln) is continuing to widen. An extreme recent example of this phenomenon is mass-scale deposition to the sequence database of proteins stemming from shotgun sequencing of microbial populations collected from seawater samples [Venter *et al.*, 2004]. In just one publication the authors of the latter publication made available 1.2 million new amino acid sequences, instantly increasing the size of the protein sequence database by roughly 50%.

Under these circumstances automated large-scale bioinformatics analysis emerges as the only available means to annotate proteins at large. Experience of the last decade shows that in each newly sequenced genome for a sizeable fraction of gene products – ranging from 30 to 70% in different species – it is possible to provide a preliminary functional and structural characterization based on similarity searches, domain finding, and structure prediction in a completely automatic fashion. Notwithstanding the well-known pitfalls of unsupervised sequence analysis [Galperin and Koonin, 1998; Bork and Bairoch, 1996], systematic first-pass annotation of completely sequenced genomes provides a useful starting point for more in-depth curated analysis.

The MIPS group (now GSF Institute for Bioinformatics, IBI) in Munich began to provide exhaustive automatic analysis of all publicly available genomes in 1996, when only five genomic sequences were published [Frishman and Mewes, 1997]. The main mission of the PEDANT genome database is to fill the gap between manually curated high quality protein sequence databases, such as Uniprot/Swissprot, and the enormous amounts of other protein sequences produced by genome sequencing projects at ever increasing pace. The PEDANT genome database is produced by systematically applying an automatic annotation pipeline to all genomic sequences that are being released in the public domain. The major premises of the PEDANT database are:

- *Timeliness.* The MIPS CPU resources make it possible to process a medium size prokaryotic genome and make it available on-line essentially overnight.
- *Completeness.* We seek to process all completely sequenced genomes as well as many incomplete genomes which are being made available by sequencing centers. In many cases PEDANT represents the only source of annotation for a given genome.
- *Standardization.* Automatic annotation of sequences follows a clearly defined protocol in terms of the particular set of bioinformatics techniques applied to each sequence and the values of pre-determined recognition thresholds used for individual methods (e.g., BLAST E-values).
- *Documentation.* Since the results of automatic sequence analyses are inevitably afflicted by a large number of false assignments we make available the raw output

of each bioinformatics method used. This allows the user to make their own judgment on the validity of functional predictions appearing on each protein's report page.

PEDANT System Architechture

The PEDANT software [Frishman et al., 2001] consists of three major parts (Figure 1): (i) the database module serves for storing, modifying and accessing data, (ii) the processing module actually carries out bioinformatics computations, and (iii) the user interface allows communication with the system through a web-based mechanism. In addition, a collection of external tools collectively referred to as 'Input module' is used for data formatting, preliminary data analysis steps (e.g. identification of genetic elements in DNA), and population of the database. Individual bioinformatics programs and the respective databases that they are using are also external to the system and have to be properly installed. The data access mechanism implemented in PEDANT is based on a standard RDBMS and the SQL language. At the present time the freely available MySQL DBMS and Oracle ™ are supported.

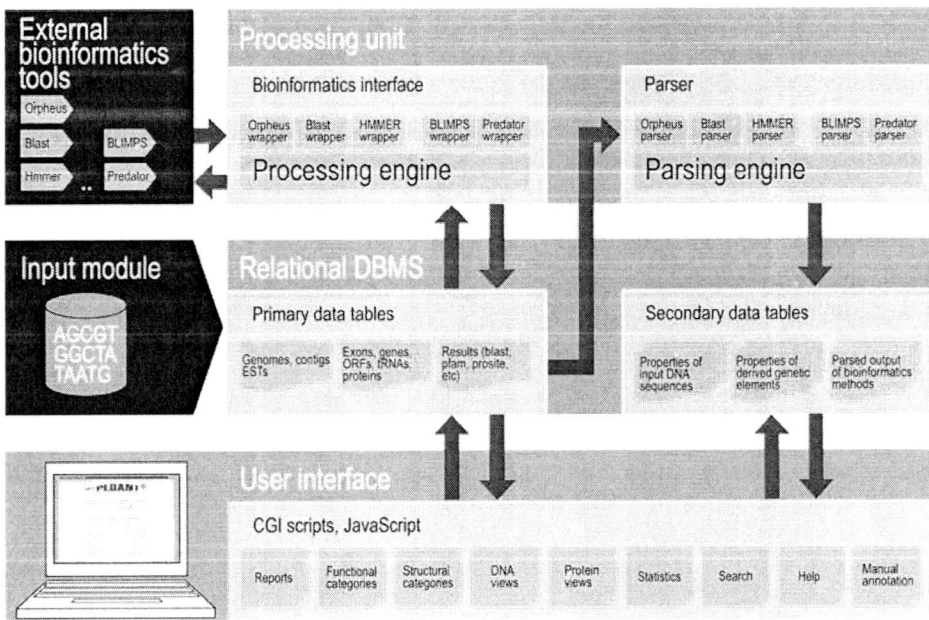

Figure 1. PEDANT architecture. The three main parts of the system are (a) relational DBMS, (b) processing unit, and (c) user interface. After pre-processing in the input module, sequence data (DNA contigs, proteins, genetic elements, exons, genes, etc.) are loaded into input primary tables. The processing unit automates the application of various bioinformatics methods (e.g. BLAST searches, secondary structure predictions) to each data element; results of the calculations are saved in the output primary tables. Results are subsequently parsed and stored in secondary tables where each piece of information (e.g. local BLAST alignments, E-values, secondary structure elements, etc.) can be individually accessed. The user interface allows accessing the data using a standard WWW browser. (Reproduced, with permission, from Frishman et al.., 2001).

The PEDANT Bioinformatics Pipeline

Since the inception of PEDANT we made every effort to keep the set of bioinformatics tools it uses as complete and up-to-date as possible. Incorporating new methods in PEDANT requires only a minimal programming effort due to its open architecture. However, we adopted a conservative approach in selecting computational methods to be included. Preference is given to widely recognized, stable, and carefully benchmarked computational tools and external databanks described in peer-reviewed publications. PEDANT maintenance thus includes systematic scanning of bioinformatics literature, testing of software tools, as well as updating already included programs as new versions and bug fixes become available.

The main vehicle for similarity searches is the PSIBLAST algorithm [Altschul *et al.*, 1997]. This method is used for general-purpose searches against the full non-redundant protein sequence databank as well as searches against a number of special datasets, including the MIPS functional categories (see below) and the COG database [Tatusov *et al.*, 2003]. In addition, detection of PROSITE [Falquet *et al.*, 2002], PFAM [Bateman *et al.*, 2004], and BLOCKS [Henikoff *et al.*, 1999] sequence motifs is performed. For those sequences that have significant matches in the Uniprot/Swissprot database, the annotation of the respective entries is analyzed and keywords and enzyme classification are extracted. Structural categorization of gene products involves PSIBLAST searches against the sequences with known 3D structure as deposited in the PDB databank [Deshpande *et al.*, 2005] and SCOP database of known structural domains [Andreeva *et al.*, 2004]. If a significant relationship exists, the secondary structure assignment of the respective three-dimensional structures as defined by the STRIDE software [Frishman and Argos, 1995] is inserted into the PEDANT structural summary in upper case. Otherwise, secondary structure information predicted by PREDATOR [Frishman and Argos, 1997] is shown in lower case. Other predicted structural features include low complexity regions [Wootton, 1994], membrane regions [Krogh *et al.*, 2001], coiled coils [Lupas *et al.*, 1991], and signal peptides [Nielsen *et al.*, 1997].

We use reasonably stringent recognition parameters to avoid excessive false positive rates, and at the same time not only provide search and prediction results in digested form, but also store the raw output of bioinformatics methods, enabling the annotator or the biologist using the database to make his own judgment on the significance of the results presented.

User Interface

The main design principle of the PEDANT user interface, which survived essentially unchanged since the very first version of the software, is the category-based mechanism of gene selection from a particular genome. This approach reflects the nature of many bioinformatics methods which allow for categorization of gene products according to some functional or structural criteria. For example, searching protein motif and family databases, such as PFAM, with appropriate search tools allows assigning functional labels to groups of proteins. Correspondingly, examples of typical PEDANT queries are "What proteins in my genome have the TIM-barrel fold, EC number 1.3.3.7, MIPS functional category 1.3.3.5, or predicted transmembrane regions?". Each such query will result in a list of gene product satisfying the search criteria. Clicking on each gene's link will produce an integrated report summarizing automatically derived annotation according to corresponding recognition thresholds. Advanced DNA and protein viewers allow visualizing the positions of genes and other genetic elements on the chromosome and predicted structural and functional information about proteins, respectively. Facilities for searching the PEDANT annotation using text queries as well as BLAST and pattern searches are provided.

Database Content

Over the past eight years the number of analyzed genomes in the PEDANT database has grown steadily (Figure 2) and stands at 371 at the time of writing, including 254 completely sequenced genomes and 117 unfinished genomic sequences from all three kingdoms of life (Figure 3). Most of these genomes were annotated in a totally unsupervised fashion. However, the database also includes several genomes that were manually annotated and, in many cases, published by MIPS. Those are *Saccharomyces cerevisiae* [Mewes *et al.*, 1997], *Thermoplasma acidophilum* [Ruepp *et al.*, 2000], *Arabidopsis thaliana* [2000], *Neurospora crassa* [Galagan *et al.*, 2003], *Parachlamydia* UWE25 [Horn *et al.*, 2004], *Listeria monocytogenes EGD*, *Listeria innocuaClip* 11262 and *H. pylori* KE26695. The total amount of data managed by PEDANT via a relational database system MySQL, is approximately 400 gigabytes, more than one gigabyte per genome on average.

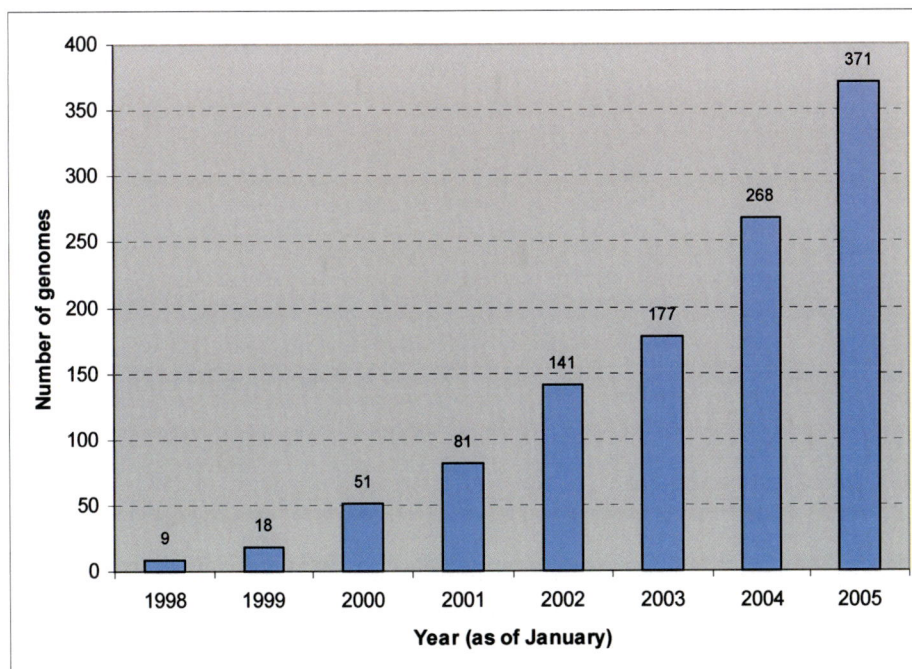

Figure 2. Growth of the number of annotated genomes in the PEDANT database since 1998. *Number as of September 1, 2004.

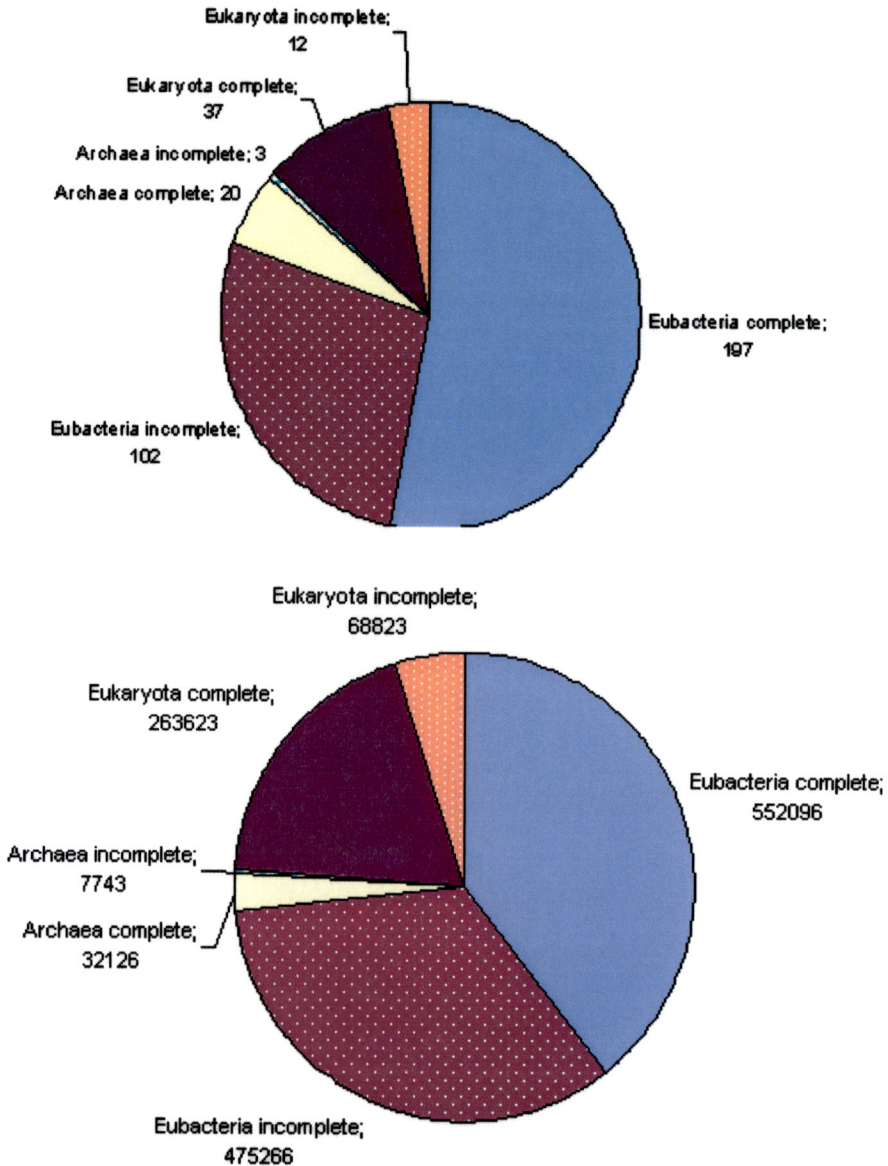

Figure 3. Number of annotated genomes (a) and protein sequences (b) in different genome categories.

To illustrate the functional and structural content of the PEDANT database we calculated the coverage of all 1 240 000 annotated protein sequences by three selected popular categories: PFAM sequence motifs, SCOP structural domains, and MIPS functional role categories [Ruepp et al., 2004]. As seen in Figure 4, the coverage varies in a wide range – from 64.3% by PFAM to 34.5% by SCOP. Only 15.2% of proteins possess all three attributes emphasizing the usefulness of applying many complementary bioinformatics techniques. The total number of all attributes computed by PEDANT for each sequence exceeds 20. The PEDANT database thus represents a valuable resource for large-scale association rule mining in automatically generated protein annotation.

Automatic Funcat

The MIPS Functional Catalogue (FunCat) was developed in 1996 and used in the annotation of *Saccaromyces cerevisiae* [Mewes *et al.*, 1997]. It comprises a hierarchically-structured classification system, which at first only contained categories describing yeast biology. Since then it has been extended and used to annotate the following genomes: *T. acidophilum, Bacillus subtilis 168, Listeria monocytogenes EGD, Listeria innocuaClip 11262, H. pylori* KE26695, *Neurospora crassa, Arabidopsis thaliana* and *Homo sapiens*. The most recent version of the FunCat (v. 2.0; [Ruepp *et al.*, 2004], is organism independent and consists of 28 main categories, covering features such as metabolism and cellular transport, as well as some more recently introduced categories (e.g. development and organ localization). The main categories are assigned a unique two-digit number e.g. 01. metabolism, which appears as the first two digits of the FunCat number. The main categories are branched into more specific categories, with up to 6 levels of increasing specificity (e.g. 01.01.06.05.01.01 biosynthesis of homocysteine).

The PEDANT software calculates automatic FunCat numbers based on a gene product's similarity to proteins in the manually annotated protein FunCat database. Although assignment of FunCat numbers by homology alone is not always reliable, it may provide useful information in the absence of manual annotation. The automatic FunCat tables for all PEDANT databases were recalculated using the new FunCat version and updated manually annotated FunCat database. Figure 5 shows the FunCat distribution of all 334 genomes in PEDANT.

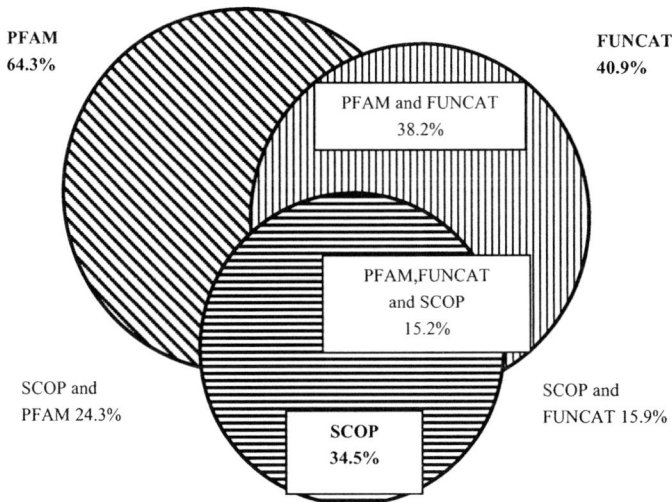

Figure 4. An illustration of the functional and structural content of the PEDANT database. The figure shows the percentage of protein sequences associated with PFAM sequence motifs, SCOP structural domains, and MIPS functional categories, as well as any combinations of these three attributes.

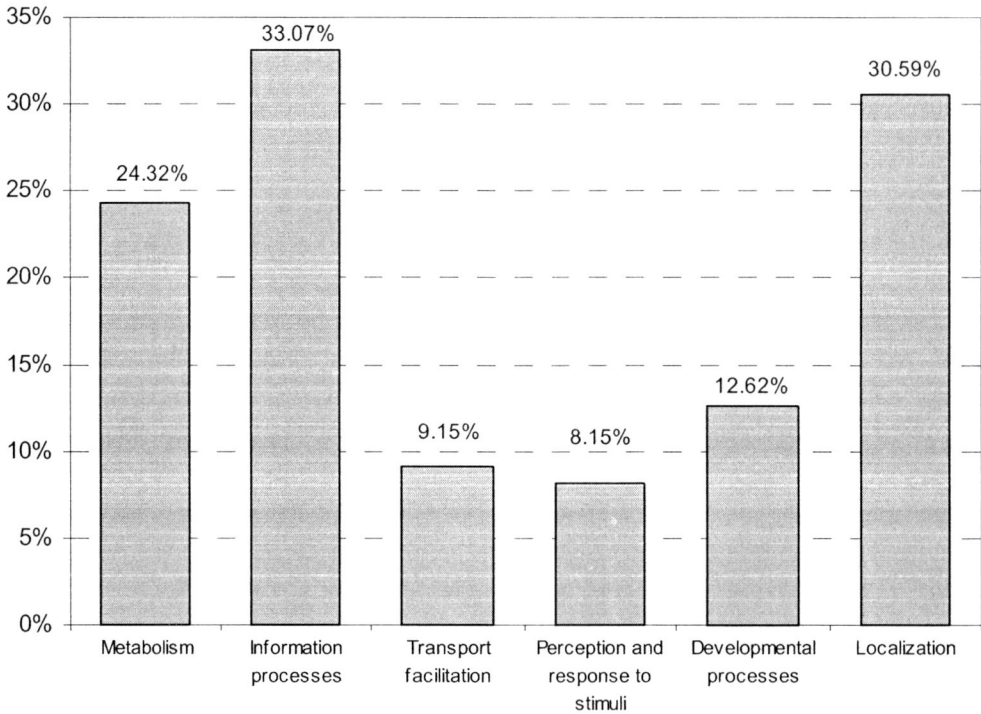

Figure 5. The FunCat distribution of all PEDANT genomes. Here the relative amounts of proteins that are assigned to one or more of the six general FunCat classes are shown. Since proteins can be assigned to more than one functional category the total fraction exceeds 100%.

Acknowledgments

We would line to thank Louise Riley and Thorsten Schmidt for valuable assistance.

References

Altschul, S.F., Madden, T.L., Schaffer, A.A., Zhang, J., Zhang, Z., Miller, W., and Lipman, D.J. (1997) Gapped BLAST and PSI-BLAST: a new generation of protein database search programs. *Nucleic Acids Res.* 25, 3389-3402.

Andreeva, A., Howorth, D., Brenner, S.E., Hubbard, T.J., Chothia, C., and Murzin, A.G. (2004) SCOP database in 2004: refinements integrate structure and sequence family data. *Nucleic Acids Res.* 32, D226-D229.

Arabidopsis Initiative (2000) Analysis of the genome sequence of the flowering plant Arabidopsis thaliana. *Nature* 408, 796-815.

Bairoch, A., Apweiler, R., Wu, C.H., Barker, W.C., Boeckmann, B., Ferro, S., Gasteiger, E., Huang, H., Lopez, R., Magrane, M., Martin, M.J., Natale, D.A., O'Donovan, C.,

Redaschi, N., and Yeh, L.S. (2005) The Universal Protein Resource (UniProt). *Nucleic Acids Res.* 33 Database Issue, D154-D159.

Bateman, A., Coin, L., Durbin, R., Finn, R.D., Hollich, V., Griffiths-Jones, S., Khanna, A., Marshall, M., Moxon, S., Sonnhammer, E.L., Studholme, D.J., Yeats, C., and Eddy, S.R. (2004) The Pfam protein families database. *Nucleic Acids Res.* 32, D138-D141.

Bork, P., and Bairoch, A. (1996) Go hunting in sequence databases but watch out for the traps. *Trends Genet.* 12, 425-427.

Deshpande, N., Addess, K.J., Bluhm, W.F., Merino-Ott, J.C., Townsend-Merino, W., Zhang, Q., Knezevich, C., Xie, L., Chen, L., Feng, Z., Green, R.K., Flippen-Anderson, J.L., Westbrook, J., Berman, H.M., and Bourne, P.E. (2005) The RCSB Protein Data Bank: a redesigned query system and relational database based on the mmCIF schema. *Nucleic Acids Res.* 33 Database Issue, D233-D237.

Falquet, L., Pagni, M., Bucher, P., Hulo, N., Sigrist, C.J., Hofmann, K., and Bairoch, A. (2002) The PROSITE database, its status in 2002. *Nucleic Acids Res.* 30, 235-238.

Frishman, D., Albermann, K., Hani, J., Heumann, K., Metanomski, A., Zollner, A., and Mewes, H.W. (2001) Functional and structural genomics using PEDANT. *Bioinformatics.* 17, 44-57.

Frishman, D., and Argos, P. (1995) Knowledge-based protein secondary structure assignment. *Proteins* 23, 566-579.

Frishman, D., and Argos, P. (1997) Seventy-five percent accuracy in protein secondary structure prediction. *Proteins* 27, 329-335.

Frishman,D. and Mewes,H.W. (1997) PEDANTic genome analysis. *Trends Genet.* 13, 415-416.

Galagan, J.E., Calvo, S.E., Borkovich, K.A., Selker, E.U., Read, N.D., Jaffe, D., FitzHugh, W., Ma, L.J., Smirnov, S., Purcell, S., Rehman, B., Elkins, T., Engels, R., Wang, S., Nielsen, C.B., Butler, J., Endrizzi, M., Qui, D., Ianakiev, P., Bell-Pedersen, D., Nelson, M.A., Werner-Washburne, M., Selitrennikoff, C.P., Kinsey, J.A., Braun, E.L., Zelter, A., Schulte, U., Kothe, G.O., Jedd, G., Mewes, W., Staben, C., Marcotte, E., Greenberg, D., Roy, A., Foley, K., Naylor, J., Stange-Thomann, N., Barrett, R., Gnerre, S., Kamal, M., Kamvysselis, M., Mauceli, E., Bielke, C., Rudd, S., Frishman, D., Krystofova, S., Rasmussen, C., Metzenberg, R.L., Perkins, D.D., Kroken, S., Cogoni, C., Macino, G., Catcheside, D., Li, W., Pratt, R.J., Osmani, S.A., DeSouza, C.P., Glass, L., Orbach, M.J., Berglund, J.A., Voelker, R., Yarden, O., Plamann, M., Seiler, S., Dunlap, J., Radford, A., Aramayo, R., Natvig, D.O., Alex, L.A., Mannhaupt, G., Ebbole, D.J., Freitag, M., Paulsen, I., Sachs, M.S., Lander, E.S., Nusbaum, C., and Birren, B. (2003) The genome sequence of the filamentous fungus Neurospora crassa. *Nature* 422, 859-868.

Galperin, M.Y., and Koonin, E.V. (1998) Sources of systematic error in functional annotation of genomes: domain rearrangement, non-orthologous gene displacement and operon disruption. *In Silico. Biol.* 1, 55-67.

Guldener, U., Munsterkotter, M., Kastenmuller, G., Strack, N., van, H.J., Lemer, C., Richelles, J., Wodak, S.J., Garcia-Martinez, J., Perez-Ortin, J.E., Michael, H., Kaps, A., Talla, E., Dujon, B., Andre, B., Souciet, J.L., De, M.J., Bon, E., Gaillardin, C., and Mewes, H.W. (2005) CYGD: the Comprehensive Yeast Genome Database. *Nucleic Acids Res.* 33 Database Issue, D364-D368.

Henikoff, S., Henikoff, J.G., and Pietrokovski, S. (1999) Blocks+: a non-redundant database of protein alignment blocks derived from multiple compilations. *Bioinformatics.* 15, 471-479.

Horn, M., Collingro, A., Schmitz-Esser, S., Beier, C.L., Purkhold, U., Fartmann, B., Brandt, P., Nyakatura, G.J., Droege, M., Frishman, D., Rattei, T., Mewes, H.W., and Wagner, M. (2004) Illuminating the evolutionary history of chlamydiae. *Science* 304, 728-730.

Keseler, I.M., Collado-Vides, J., Gama-Castro, S., Ingraham, J., Paley, S., Paulsen, I.T., Peralta-Gil, M., and Karp, P.D. (2005) EcoCyc: a comprehensive database resource for Escherichia coli. *Nucleic Acids Res.* 33 Database Issue, D334-D337.

Krogh, A., Larsson, B., von, H.G., and Sonnhammer, E.L. (2001) Predicting transmembrane protein topology with a hidden Markov model: application to complete genomes. *J. Mol. Biol.* 305, 567-580.

Lupas, A., Van, D.M., and Stock, J. (1991) Predicting coiled coils from protein sequences. *Science* 252, 1162-1164.

Mewes, H.W., Albermann, K., Bahr, M., Frishman, D., Gleissner, A., Hani, J., Heumann, K., Kleine, K., Maierl, A., Oliver, S.G., Pfeiffer, F., and Zollner, A. (1997) Overview of the yeast genome. *Nature* 387, 7-65.

Nielsen, H., Engelbrecht, J., Brunak, S., and von, H.G. (1997) Identification of prokaryotic and eukaryotic signal peptides and prediction of their cleavage sites. *Protein Eng* 10, 1-6.

Pagel, P., Kovac, S., Oesterheld, M., Brauner, B., Dunger-Kaltenbach, I., Frishman, G., Montrone, C., Mark, P., Stumpflen, V., Mewes, H.W., Ruepp, A., and Frishman, D. (2004) The MIPS mammalian protein-protein interaction database. *Bioinformatics.*

Ruepp, A., Graml, W., Santos-Martinez, M.L., Koretke, K.K., Volker, C., Mewes, H.W., Frishman, D., Stocker, S., Lupas, A.N., and Baumeister, W. (2000) The genome sequence of the thermoacidophilic scavenger Thermoplasma acidophilum. *Nature* 407, 508-513.

Ruepp, A., Zollner, A., Maier, D., Albermann, K., Hani, J., Mokrejs, M., Tetko, I., Guldener, U., Mannhaupt, G., Munsterkotter, M., and Mewes, H.W. (2004) The FunCat, a functional annotation scheme for systematic classification of proteins from whole genomes. *Nucleic Acids Res.* 32, 5539-5545.

Tatusov, R.L., Fedorova, N.D., Jackson, J.D., Jacobs, A.R., Kiryutin, B., Koonin, E.V., Krylov, D.M., Mazumder, R., Mekhedov, S.L., Nikolskaya, A.N., Rao, B.S., Smirnov, S., Sverdlov, A.V., Vasudevan, S., Wolf, Y.I., Yin, J.J., and Natale, D.A. (2003) The COG database: an updated version includes eukaryotes. *BMC. Bioinformatics.* 4, 41.

Venter, J.C., Remington, K., Heidelberg, J.F., Halpern, A.L., Rusch, D., Eisen, J.A., Wu, D., Paulsen, I., Nelson, K.E., Nelson, W., Fouts, D.E., Levy, S., Knap, A.H., Lomas, M.W., Nealson, K., White, O., Peterson, J., Hoffman, J., Parsons, R., Baden-Tillson, H., Pfannkoch, C., Rogers, Y.H., and Smith, H.O. (2004) Environmental genome shotgun sequencing of the Sargasso Sea. *Science* 304, 66-74.

Wootton, J.C. (1994) Non-globular domains in protein sequences: automated segmentation using complexity measures. *Comput. Chem.* 18, 269-285.

In: In Silico Genomics and Proteomics
Editors: N. Mulder and R. Apweiler, pp. 157-183

ISBN 1-59454-995-8
© 2006 Nova Science Publishers, Inc.

Chapter 12

Microbial Genomes at NCBI

William Klimke and Tatiana Tatusova
National Center for Biotechnology Information,
National Library of Medicine, Building 38A Bethesda, MD 20894

Abstract

An ever increasing number of complete and unfinished microbial genomes are being submitted to the public databases. Proper genome analysis is of paramount importance in understanding genomic biology and in annotating genome sequences. The National Center for Biotechnology Information (NCBI) provides a number of resources to analyze microbial genomes and proteomes. These include various databases housing the original data submissions, as well as explicit links connecting them. Precalculated results for various similarity searches such as sequence, domain, and structural, are stored to facilitate quick retrieval. Data from both the original submission and precalculated results are browsable and retrievable through the various web portals and displays available at NCBI and also through the FTP site or through the set of programs collectively known as E-Utilities. A set of interactive tools provides additional access to a number of precalculated similarity results that allow users to manipulate the results in a biologically meaningful way. All of these resources provide a rich set of tools for the analysis and understanding of genomic biology.

Introduction

As this chapter was being written the 218th complete genome was deposited to the public databases, and it is likely that this number will have increased significantly once this book is published. The scale of genome sequencing and the production of data has reached astounding proportions since the completion of the first microbial genome (*Haemophilus influenzae* Rd KW20) was released in 1995 and both the number of genomes and the number of unique genera for which a completely sequenced genome is available are increasing rapidly (Figure 1; Fleischmann *et al.*, 1995).

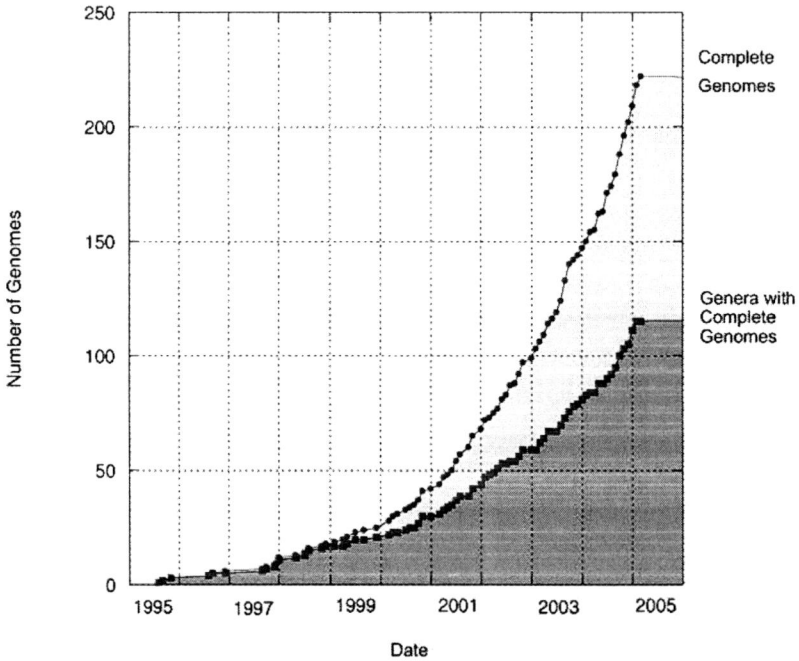

Figure 1. Growth of complete microbial genomes in the last ten years. The number of completed microbial genomes (green) and the number of genera (blue) for which a complete genome is available are shown. Information is based on the date when complete sequence data was publicly available. The number of completed genomes (218) has increased significantly since the publication of the first complete microbial genome sequence in 1995 by The Institute for Genomic Research (TIGR) for *Haemophilus influenzae* Rd KW20 (Fleischmann *et al.*, 1995). The first 100 genomes were achieved in eight years at the beginning of 2003, while the second 100 genomes were achieved in less than two years, at the end of 2004. A large proportion of sequencing projects involve sequencing closely related strains or species for comparative genomics so that there are only 115 genera for which a complete genome is available.

The number of proteins encoded by these microbial genomes alone exceeds 650 000. The 218 complete genomes includes 21 archaeal and 197 bacterial genomes representing a wide range of organisms (Table 1). They include many important human pathogens, but also organisms that are of interest for non-medical reasons. There are obligate intracellular parasites, symbionts, free-living microbes, hyperthermophiles and psychrophiles, and aquatic and terrestrial microbes, all of which have provided a rich insight into evolution and microbial biology and ecology. There is almost a 20-fold range of genome sizes, spanning from 490 Kbp for the archaeal parasite *Nanoarchaeum equitans* Kin4-M to over 9 Mbp for the soil-dwelling filamentous antibiotic-producing *Streptomyces* (*Streptomyces coelicolor* A3(2) and *Streptomyces avermitilis* MA-4680) and the nitrogen-fixing symbiotic microbe *Bradyrhizobium japonicum* USDA 110 (Bentley *et al.*, 2002; Ikeda *et al.*, 2003; Kaneko *et al.*, 2003; Waters *et al.*, 2003). There are organisms with single circular chromosomes, but also organisms with linear chromosomes, multiple chromosomes, and a mixture of chromosomes and extrachromosomal elements including plasmids. The Lyme disease-causing spirochaete *Borrelia burgdorferi* contains the largest number of extrachromosomal elements with 21 small circular and linear plasmids along with the linear chromosome comprising the complete genome (Fraser *et al.*, 1997). The GC-content of bacterial genomes also spans a large range, from 22.5% for the obligate intracellular symbiotic microbe, *Wigglesworthia glossinidia*, to 72% for the above-mentioned *Streptomcyes* genomes (Akman *et al.*, 2002). In addition to the 218 completed genome sequences, many more sequencing projects are in an unfinished state and have incomplete data submitted or do not have any sequence data at all (see below). Several hundred projects are listed at GOLD (genomes online database) and it is obvious that sequence data will continue to grow (Bernal *et al.*, 2001).

Table 1. Complete Microbial Genomes by Major Taxonomic Group[a]

Major Taxonomic Group[b]	Number of[c]		
	Genus[d]	Species[e]	Genomes[f]
Complete			
All Prokaryotes	115	178	218
Archaea[g]	16	21	21
Crenarchaeota	3	4	4
Euryarchaeota	12	16	16
Nanoarchaeota	1	1	1
Bacteria[g]	99	157	197
Actinobacteria	9	16	17
Aquificae	1	1	1
Bacteroidetes/Chlorobi	3	4	4
Chlamydiae	3	5	8
Chloroflexi	1	1	1
Cyanobacteria	6	7	9
Deinococcus/Thermus	2	2	3
Firmicutes	15	40	58
Mollicutes	4	11	11
Bacillales	5	13	26
Lactobacillales	4	12	17
Clostridia	2	4	4
Fusobacteria	1	1	1
Proteobacteria	53	73	87
Alpha	15	18	20
Beta	7	10	11
Gamma	24	37	46
Delta	4	4	4
Epsilon	3	4	6
Planctomycetes	1	1	1
Spirochaetes	3	5	6
Thermotogae	1	1	1

[a]. Complete genomes publicly available as of Jan. 31/2005.
[b]. Major taxa for which complete genomes are available.
[c]. Unique members for each category for which complete genomes are available.
[d]. Number of unique genera that have complete genomes per taxonomic group.
[e]. Number of unique species that have complete genomes per taxonomic group.
[f]. Total number of genomes per taxonomic group.
[g]. The two major branches of prokaryotes are underlines
[h]. Two taxa with large numbers of representatives are split into subgroups at the class level.

No matter how impressive the numbers of genome sequencing projects are, they represent a miniscule fraction of the total number of bacterial species that are potential sequencing targets. Recent quantitative fluorescent *in situ* hybridization analyses using 16S rRNA probes indicate that there are approximately 160 different taxa per ml of seawater, while there are between 6400 and 38 400 different taxa in a single gram of soil, and the total number of prokaryotic species worldwide is estimated to be between 10^7 and 10^9 (Curtis *et al.*, 2002; Dykhuizen, 1998). There are currently 52 phyla in the prokaryotic tree, 26 of which have cultured representatives and 26 that do not (Rappe and Giovannoni, 2003). The cultivated subset of bacterial species represents a mere 4500 or so organisms, most from a small number of phyla, that preferentially grow on agar plates. Obviously the cultured subset skews our understanding of microbial life and more importantly the biochemical processes that have a global impact on the planet (Torsvik *et al.*, 2002). Recently, a new field of genomics, metagenomics, has been initiated to overcome the limitations of culturing organisms prior to sequencing (Handelsman, 2004). Sequencing projects from an acid-mine drainage site, the Sargasso Sea, and deep-sea methane seeps have recently been completed (Hallam *et al.*, 2004; Tyson *et al.*, 2004; Venter *et al.*, 2004). Although the field of metagenomics will allow new insight into microbial processes, it is necessarily complicated by the fact that reconstruction of complete genomes is limited by the type of data generated. The result is that future genomic analysis tools will have to take into consideration the uncertain origin of the DNA sequences during analysis. Making sense of genomic data as well as incorporating these analyses into future genome submissions is one goal that is aided by the various databases and tools present at NCBI.

Data Organization

The primary mission of NCBI is to develop information systems for molecular biology and the institute has done this through the establishment of a number of searchable databases and analysis tools (URLs for all databases and tools can be found in Table 3). The Entrez database system consists of a set of nodes that contain primary data submissions with an integrated and searchable mechanism that allows quick retrieval of data (Wheeler *et al.*, 2005). Submissions can be nucleotide sequences submitted to the International Nucleotide Sequence Data collaboration, which is a joint cooperative effort between GenBank at NCBI, and DDBJ (DNA Data Bank of Japan), and EMBL-EBI (European Bioinformatics Institute), protein sequence derived from translation of the deposited nucleotide sequences or from other database sources such as SWISS-PROT (now UniProt) and the Protein Data Bank (PDB), microarray datasets and profiles in the Gene Expression Omnibus (GEO), publications from PubMed and PubMed Central that include the abstracts from MEDLINE, and organism names in the taxonomy database that contains over 165 000 organism names that are represented by at least sequence, whether it be nucleotide or protein (Barrett *et al.*, 2005; Benson *et al.*, 2005; Kanz *et al.*, 2005; Miyazaki *et al.*, 2004; Wheeler et al., 2005). All entries are linked by indexing the information present in each record and by connecting to the other Entrez databases, which can be accessed by the link menu available on each record.

Table 2. Databases relevant to microbial genomes[a]

Database[b]	Data Stored	Description
Entrez		
Gene	gene	gene-related information
Genome	genome	genomes and maps from all organisms
Genome Project	project	organism-specific genomic projects
Nucleotide	nucleotide	DNA sequences including genomes
Protein	protein	proteins derived from nucleotide translations and third-party databases
literature	publications	publications in MEDLINE, PMC, etc.
Structure	3D structures	structural information from Molecular Modeling Database (MMDB)
CDD	conserved domains	conserved protein domains from a number of publicly available databases
3D Domains	3D domains	compact domains identified automatically in MMDB
Taxonomy	taxonomic	organism names and taxonomic hierarchies
Other		
Trace Archive	raw sequences	raw sequence reads generated by sequencing projects
Assembly Archive	assembly and alignments	assembly information and alignments of the traces present in the Trace Archive
COG	orthologous protein clusters	clusters of proteins from at least 3 phylogenetic lineages
RefSeq	Curated sequences	comprehensive, integrated, non-redundant set of curated sequences for genomes

[a] Other Entrez databases are shown in Table 3 and are available through the web

[b] . Databases shown in bold contain primary data while the others contain derived data

Table 3. URLs of NCBI databases, tools, and documentation

Database/Tool/Site	URL
NCBI	http://www.ncbi.nlm.nih.gov
Entrez	http://www.ncbi.nlm.nih.gov/Entrez/
CDD	http://www.ncbi.nlm.nih.gov/entrez/query.fcgi?db=cdd
Genome	http://www.ncbi.nlm.nih.gov/entrez/query.fcgi?db=Genome
Genome Project	http://www.ncbi.nlm.nih.gov/entrez/query.fcgi?db=genomeprj
Gene	http://www.ncbi.nlm.nih.gov/entrez/query.fcgi?db=gene
Nucleotide	http://www.ncbi.nlm.nih.gov/entrez/query.fcgi?db=Nucleotide
Protein	http://www.ncbi.nlm.nih.gov/entrez/query.fcgi?db=Protein
PubMed	http://www.ncbi.nlm.nih.gov/entrez/query.fcgi?db=PubMed
PubMed Central	http://www.ncbi.nlm.nih.gov/entrez/query.fcgi?db=pmc
Structure	http://www.ncbi.nlm.nih.gov/entrez/query.fcgi?db=Domains
Taxonomy	http://www.ncbi.nlm.nih.gov/entrez/query.fcgi?db=Taxonomy
Other NCBI Databases	
AssemblyArchive	http://www.ncbi.nlm.nih.gov/projects/assembly
COGs	http://www.ncbi.nlm.nih.gov/COG/
RefSeq	http://www.ncbi.nlm.nih.gov/RefSeq/
TPA	http://www.ncbi.nih.gov/Genbank/TPA.html
TraceArchive	http://www.ncbi.nlm.nih.gov/Traces/trace.cgi?
FTP	ftp://ftp.ncbi.nih.gov/
GenBank Release	ftp://ftp.ncbi.nih.gov/genbank/
GenBank Genomes	ftp://ftp.ncbi.nih.gov/genbank/genomes/Bacteria/
RefSeq Release	ftp://ftp.ncbi.nih.gov/refseq/
RefSeq Genomes	ftp://ftp.ncbi.nih.gov/genomes/Bacteria/
BLAST	http://www.ncbi.nlm.nih.gov/BLAST/
Genomic BLAST	http://www.ncbi.nlm.nih.gov/sutils/genom_table.cgi
Genomic BLAST[1]	http://www.ncbi.nlm.nih.gov/genomes/geblast.cgi?

Table 3. URLs of NCBI databases, tools, and documentation (continued)

Database/Tool/Site	URL
Precomputed tools[1]	
COGTable	http://www.ncbi.nlm.nih.gov//sutils/coxik.cgi?
GenePlot	http://www.ncbi.nlm.nih.gov/sutils/geneplot.cgi?
gMap	http://www.ncbi.nlm.nih.gov/sutils/gmapi?
TaxPlot	http://www.ncbi.nlm.nih.gov/sutils/taxik2.cgi
TaxTable (TaxMap)	http://www.ncbi.nlm.nih.gov/sutils/taxik.cgi?
CDDTable	http://www.ncbi.nlm.nih.gov/sutils/genomesCDD.cgi?
PDBTable	http://www.ncbi.nlm.nih.gov/sutils/tablik.cgi?
Other NCBI Info	
Bacillus anthracis info	http://www.ncbi.nlm.nih.gov/genomes/MICROBES/anthracis.html
EUtilities	http://eutils.ncbi.nlm.nih.gov/entrez/query/static/eutils_help.html
Genomic Biology	http://www.ncbi.nlm.nih.gov/Genomes/index.html
NCBI Handbook	http://www.ncbi.nlm.nih.gov/books/bv.fcgi?call=bv.View..ShowTOC&rid=handbook.TOC&depth=2
NCBI Education	http://www.ncbi.nlm.nih.gov/Education/index.html
Prokaryotic Table	http://www.ncbi.nlm.nih.gov/genomes/lproks.cgi
PDBTable	http://www.ncbi.nlm.nih.gov/sutils/static/PDB_bact.html

[1] These tools require the input of the Entrez Genome UID in order to function. Links to all tools are provided in Entrez Genome and Entrez GenomeProject. The UID format should be'' .cgi?gi=UID''

For example, the taxonomic information and publication record for a given nucleotide sequence are cross referenced with the NCBI taxonomy and PubMed databases, respectively, while the protein sequences encoded on this entry are stored in the protein database. This interconnection now links over 20 databases, and a new Global Query allows searching across all databases simultaneously.

In addition to the explicit links that come from indexing the information in each record, a number of precomputed results are stored in a separate database which allows rapid retrieval. These include precomputed sequence similarity results that are discussed in the analysis section of this chapter, and publication similarity searches that retrieve a list of citations that are closely related to the article in question. Besides the publications that were present on the original record or calculated by similarity, important curated publication links are stored in specific databases as well, including domain-specific publications at CDD. A new system initiated for Entrez Gene, the Gene Reference Into Function (GeneRIFs), allows outside submitters and NCBI indexers to link publications that were not on the original submission, to Genes records, thus providing a way to enhance the descriptions and annotations of protein function found in the Gene and Genome entries (Maglott *et al.*, 2005). This entire system enables finding associations that are not explicitly stated in the original data submission. For example, the pre-computed BLAST results from a given protein encoded by a genome can be used to find a related SWISS-PROT record that contains annotated information and publications not found on the original genome. New associations provides a rich resource for biologists to mine for important data that allows increased understanding of genomic biology.

The list of databases and resources that are relevant to microbial genomics is shown in Table 2 and includes databases both inside of the Entrez system and outside as well as specific analysis tools that are discussed below. These include RefSeq genomic records, which are derived from the data in GenBank and that aim to provide a collection of manually curated, integrated, non-redundant, up-to-date information that represents a synthesis of annotation for proteins and genes (Pruitt, 2005). In addition to the Entrez databases, there are a number of other databases that contain information relevant to microbiologists. These include the Cluster of Orthologous Groups (COGs) that will be discussed in the analysis section. There are also the Trace Archive that contains the raw sequence reads from genome sequencing projects and the Assembly Archive that combines the traces with the assembly information present in GenBank (Salzberg *et al.*, 2004). The Trace Archive captures all of the sequence reads generated during a sequencing project, while the Assembly Archive records the alignments of the traces that was used to generate the final genome sequence. The Assembly Viewer provides a graphical tool for visualization of the multiple alignments of the trace sequences. These two databases provide a way to scrutinize the polymorphisms found between two or more genomes by analyzing the underlying data used to produce the genomic sequence. The initial set of genomic entries in the Assembly Archive consisted of seven closely related strains of *Bacillus anthracis* that were sequenced following the anthrax bioterror attacks in 2001 that resulted in four high-quality SNPs (single nucleotide polymorphisms) between the two sequenced chromosomes (Read *et al.*, 2002). The low number of SNPs necessitated analyzing the sequence traces underlying those SNPs with a great deal of scrutiny and led to the establishment of the Assembly Archive. Mining all of these databases generates a great deal of data that can be used for the understanding of microbial genomics.

Besides the internal links connecting NCBI-specific database records, NCBI provides a number of external links to third-party databases. Examples include the Enzyme Commission (EC) numbers which have been assigned by the Nomenclature Committee of the International Union of Biochemistry and Molecular Biology (IUBMB) to the Enzyme nomenclature database at Expasy and the American Type Culture Collection (ATCC) numbers assigned to microbial strains (Bairoch, 2000). Links are provided as both the database cross references, for example in protein and gene records, and also through the LinkOut functionality. Additional links such as those for gene ontology, KEGG (Kyoto Encyclopedia of Genes and Genomes) pathways, and species-specific databases such as EcoCyc for *E. coli* are also in the process of being added (Kanehisa *et al.*, 2004; Kessler *et al.*, 2005; Lewis, 2005). Protein interactions from the BIND (Biomolecular Interaction Network Database) database are available in Entrez Gene and NCBI is working towards providing additional third-party links (Alfarano *et al.*, 2005).

Genome Project Database

Although the existing Entrez databases make it easy to store, manipulate, and query genomic data, they do not facilitate the organism- and genome-centric view that most biologists think of when they consider a complete microbial genome. Towards that end NCBI created the Genome Project database which consists of organism-specific overviews that function as portals from which all projects in the database pertaining to that organism can be browsed and retrieved, and from which all data, including nucleotide, genomic, protein, and publications, can be easily obtained. This database consists of projects for which sequence data has been deposited and also contains projects that have been publicly listed or have received funding from public agencies but have not yet deposited any sequences. The organism-specific portal allows users to find not only the genome sequencing project, but also to easily find related ones such as gene expression project for that organism that has microarray data deposited in GEO. Specific information pertaining to the importance of each sequencing project and the reason for sequencing each organism has been attached to each project by NCBI curators, which is particularly useful as this information is almost never found on the original genome submission. A special effort has been made to provide important strain-specific information that notes whether the microbial strain in question is a clinical isolate, an environmental isolate, or a laboratory strain, and each Genome Project links to related projects (strains) of the same species if any are present. Physiological attributes for each organism are attached to the respective genome project. The information from all projects, including genome information and the physiological attributes, is collated and present in the Prokaryotic Table that lists all complete and in progress Genome Projects (Figures 2 and 3). The Prokaryotic Table provides a portal to both the Genome Project and the Genome database. The table consists of three tabs, two tabs based on sequence availability and the third based on the organism attributes described above.

A.

***Bacillus anthracis* str. Ames Ancestor**. This is the type strain (0581, A2084, genotype 62, Group A3.b) for *Bacillus anthracis* and contains the two virulence plasmids, pOX1 and pOX2, that encode anthrax toxin and capsule, respectively, making this a virulent strain. This strain is considered the "gold standard" for *B. anthracis*.

Cellular features					Environment			Temperature	
Gram stain	Shape	Arrangement	Endospores	Motility	Salinity	Oxygen Req.	Habitat	Opt. temp.	Range
+	Rod	Singles, Pairs, Chains	Yes	Yes		Facultative	Terrestrial		Mesophilic

Pathogenic in: Animal **Disease:** Anthrax

B.

organism group: [Firmicutes ▾]

* size is estimated, otherwise genome size is calculated based on existing sequences

sequencing status filter: ○ all; ⊙ complete; ○ assembly; ○ no sequence.

Legend: ▢ - complete; ▢ - assembly; ▢ - no sequence.

[2] **79 Microbial Genomes selected: Complete - 79, Assembly - 0, Unfinished - 0**

								clear filter				save	
Organism	**Kng**	* **Size**	**GC**	**Gram stain**	**Shape**	**Arrangement**	**Endospores**	**Motility**	**Oxygen Req.**	**Habitat**	**Temp. range**	**Pathogenic in**	**Disease**
Aster yellows witches'-broom phytoplasma AYWB	B	0.72	26.8		Sphere				Aerobic	Host-associated	Mesophilic	Plants	aster yellows, witches'-broom
Bacillus anthracis str. 'Ames Ancestor'	B	5.5	35.2	+	Rod	Singles, Pairs, Chains	Yes	Yes	Facultative	Terrestrial	Mesophilic	Animal	Anthrax
Bacillus anthracis str. Ames	B	5.23	35	+	Rod	Singles, Pairs, Chains	Yes	Yes	Facultative	Multiple	Mesophilic	Animal	Anthrax
Bacillus anthracis str. Sterne	B	5.23	35.4	+	Rod	Singles, Pairs, Chains	Yes	Yes	Facultative	Multiple	Mesophilic	Animal	Anthrax

Organism info   Complete genomes   Genomes in progress

Figure 2. Genome Project and Prokaryotic Table view of *Bacillus anthracis*. Genome Projects collect all sequence information for a given organism and present it in an overview display. Links to sequencing information, precomputed results, and related projects at NCBI are provided as well as external links to the sequencing center and other sites that contain information of interest to this particular organism (not shown). In addition, an organism description highlights why this organism was sequenced and provides strain-specific information that is useful in distinguishing related strains. Physiological attributes are provided and include information on cellular attributes (gram stain, shape, arrangement), and whether the organism produces endospores and is motile. It also includes temperature information (ranges and optimal temperature), environmental and metabolic information (habitat, oxygen requirements, and salinity), and whether the organism is a pathogen and, if so, what diseases it causes. Information in the attribute table has been collected by curators from the literature and not all fields have been populated. The prokaryotic table provides three views for Genome Projects, two based on sequence availability (complete genomes and genomes in progress) and one that collates the attribute information from each project (organism info). All three tabs are filterable by sequence availability and major taxonomic groups. Each individual column can be sorted in ascending or descending order, and the output from the entire table can be saved in a text file. The organism name is a direct link to the genome project. A. The strain-specific description and attribute table from the *Bacillus anthracis* str. 'Ames Ancestor' Genome Project is shown. B. Organism info tab of the Prokaryotic Table showing a filtered set of complete genomes for all firmicutes and sorted by organism names. The set of 3 complete *Bacillus anthracis* genome projects are near the top.

A.

Chromosomes: *chromosome*
Plasmids: pXO1, pXO2

Genome Info:	Features:	BLAST homologs:	Links:	Review Info:
Refseq: NC_007530	Genes: 5635	COG	Genome Project	Publications
GenBank: AE017334	Protein coding: 5309	3D Structure	Refseq FTP	Refseq Status: **Provisional**
Length: 5,227,419 nt	Structural RNAs: 128	TaxMap	GenBank FTP	Seq.Status: **Completed**
GC Content: 35%	Pseudo genes: 1	TaxPlot	BLAST	Sequencing center: TIGR
% Coding: 80%	Others: 6	GenePlot	TraceAssembly	Completed: 2004/05/20
Topology: circular	Contigs: 1	gMap	CDD	Organism Group
Molecule: DNA			Other genomes for species	

Gene Classification based on COG functional categories!

Search gene, GeneID or locus_tag:

[Find Gene]

Zoom

B.

size is estimated, otherwise genome size is calculated based on existing sequence
79 Complete Microbial Genomes selected: [A] - 0, [B] - 79

clear filter   save

Tools legend: T - TaxTable; P - ProtTable; C - COG Table; B - 3-D neighbor; L - BLAST; S - CDD search; F - FTP; R - Publication.

F	Organism	King	Group	Size	GC	relur	nplsm	GenBank	RefSeq	Released	Center	Tools
	Acinetobacter sp. ADP1	B	Firmicutes	0.72	26.8		4	CP000061	NC_007716	03/13/2006	Ohio State University	T P C B L S R
	Bacillus anthracis str. 'Ames Ancestor'	B	Firmicutes	5.5	35.2		2	AE017334	NC_007530	05/20/2004	TIGR	T P C B L S R
	Bacillus anthracis str. Ames	B	Firmicutes	5.23	35		1	AE016879	NC_003997	05/07/2003	TIGR	T P C B L S R
	Bacillus anthracis str. Sterne	B	Firmicutes	5.23	35.4		1	AE017225	NC_005945	06/24/2004	DOE Joint Genome Institute	T P C B L S R

Figure 3. Genome view and Prokaryotic Table for *Bacillus anthracis*. A. Genome view of *Bacillus anthracis* str. 'Ames Ancestor'. The genome view displays all chromosomes and plasmids that constitute the genome for that organism. Links are provided for both the GenBank and the RefSeq versions of the sequence as well as all of the precalculated tables described in the analysis section (protein, structural RNAs, COGTable, PDBTable, TaxTable, TaxPlot, GenePlot, CDD Table). Links are also provided to both FTP sites, the genomic BLAST pages, and to the Trace Assembly Archive if sequence from this organism is present in that database. Below the table is a map view showing a zoomed in view near the start site of the sequence with genes color-coded according to COG functional categories. The zoom level and the map view can be altered by the controls above the linear map. The circular map to the right shows the position in the chromosome/plasmid under examination and it can be used to find a specific gene from this organism. A tooltip pops up when the mouse cursor is over a gene showing the gene name, locus_tag, protein name, COG group, and location. Clicking on a gene shows links to Entrez Protein and Entrez Gene. B. The same set of genomes highlighted in Figure 2B are shown in the complete genomes tab. This tab collates the information in the genome view and displays it, listing Accession Numbers and the precalculated tables present in panel A as well as the release date, sequencing center, and publication links. This table is also filterable and sortable and has the same settings as the table in panel 2B. Not shown is the in progress tab that displays alternative information to that in C. The number of contigs and genome size (calculated from the deposited sequence, or estimated based on the literature), links to the specific BLAST page and sequencing centers are available. There are no pre-computed results such as the COGTable for incomplete genomes.

Each tab is filterable by sequence availability and by major taxonomic groups, and many of the columns are sortable, allowing users to limit the range of organisms for analysis.

The Genome Project database thus provides a mix of data: data which is typically associated with genome sequencing projects such as nucleotides and proteins along with more organism-specific data such as ecological niche, medical or environmental importance, and physiological attributes, all of which are usually not found on genome submissions, to provide a richer resource for understanding microbial genomes in terms of biological context.

Data Representation and Retrieval

The Entrez system provides a rich set of search tools for the retrieval and display of information in the various databases, which can be done through the web interface. Information can be browsed or searched, and a specific record or set of records can be found by using Boolean queries, filters, and limits. The aforementioned links system is found in the upper right corner of each display. Individual or groups of records can be displayed or downloaded in a variety of formats including the typical flat file view, but also in others including, but not limited to, ASN.1 (Abstract Syntax Notation 1), FASTA (sequence), and XML (Extensible Markup Language). GenBank and RefSeq releases contain all of the sequence information pertinent to genomic records. Genome-specific information is kept in the Genomes FTP, and of particular interest for the study of microbial genomes are the additional tables including the Protein, RNA, and PDBTables which list all proteins, RNAs, and precalculated structural similarities, respectively. In addition, some of the Genomes FTP directories contain the results of both Glimmer and GeneMark genefinding algorithms for that particular genome (Delcher *et al.*, 2001; Besmer *et al.*, 2001). Precalculated results for similarity to known structures, COGs, and BLAST similarity searches are also available and will be discussed below. Gene-specific information, including publications attached specifically to that gene as opposed to the entire genome (GeneRIFs) are available in Entrez Gene (Maglott *et al.*, 2005). Protein-protein interactions will also be linked from Entrez Gene as has already been done for the HIV protein interaction database.

There are also a number of views built for browsing the genome. These include the genome view available by clicking on the genome map present on each display page in Entrez Genome. This display allows one to graphically view a section of the genome with the genes color-coded according to COG functional category, and searches by gene name can be initiated. Investigation into additional open reading frames can be initiated using NCBI's ORF Finder. There is also a link to the graphical view that is available from the genome display page and via the nucleotide record.

The Entrez webview provides a simple way to browse and retrieve one or a handful or records, while the FTP directories provide ways of downloading large amounts of data. The third option is the Entrez Programming Utilities (E-Utilities) function which provides a structured interface to the Entrez system. It consists of a set of seven server-side programs that provide a stable interface to search, retrieve, and link records found in Entrez using a fixed URL to translate a standard set of input parameters into the values necessary for various NCBI software components to search for and retrieve the requested data. Users can utilize a number of programming languages to send URLs to the eUtils server and interpret the XML response for customized retrieval and downstream data processing. These utilities provide users with the ability to build customized data pipelines for the analysis of genomic data.

Data Analysis

Sequence Similarity

One of the first steps undertaken when generating accurate genome annotation or when analyzing genomic data is when protein sequences encoded by a given genome are used to query the database of existing proteins for sequence similarity. Proteins with a high degree of similarity across the entire length of the protein are likely to perform the same function and if that function is known, transfer of annotation can be done automatically. Proteins with no, or very low, similarity to proteins of known function are typically referred to as hypothetical or unknown proteins. The set of proteins that exhibit weak similarity, or where similarity does not extend the entire length of the protein, may prove recalcitrant to analysis and require manual analysis and annotation. In order to facilitate sequence similarity analyses, NCBI has built a number of tools for both on-the-fly similarity search as well as a number of precalculated search results that are available as links from the various databases. Although sequence similarity searching is a powerful tool and a useful first step in genome analysis, it does have limitations, and a more thorough analysis may be required before a function or annotation is applied to a given protein sequence.

BLAST

The Basic Local Alignment Search Tool (BLAST) set of programs allows a variety of sequence-similarity searches against a number of the NCBI sequence databases including the trace archive (Altschul *et al.*, 1990; Altschul *et al.*, 1997; Mcginnis and Madden, 2004). All of these programs are available through the NCBI BLAST page or can be downloaded for in-house BLAST searches using existing or custom-built databases.

Genomic and Environmental Blast

The genomic BLAST pages provides a way to query either the nucleotide sequence or all of the proteins encoded by a single organism or by taxonomic group. (Cummings *et al.*, 2002). Nucleotide searches of the genomic BLAST pages utilize MegaBLAST which handles queries more rapidly than normal BLASTN. Microbial genomes are searchable using genomic BLAST, including those for which sequence data is not yet publicly available through Entrez. The large number of sequences derived from both the Sargasso Sea (over 1 billion Bp) and the Acid Mine Drainage (76 million Bp) and the over 1 million proteins encoded by these sequences prompted the creation of a new BLAST database for searching against these environmental sets. Searches limited to the env_nt (contigs) or the env_nr (proteins) databases from the main BLAST pages will query the sequences derived from these metagenomics studies.

gMap

Genomic sequence comparison is becoming a more common tool to understanding genome evolution as more genomes are sequenced, and this is especially true for closely related strains. The largest groups of species for which some sequence data is available include *Bacillus anthracis* (10), *Escherichia coli* (6), *Haemophilus influenzae* (6), and *Streptococcus pyogenes* (6), *Salmonella enterica* (6), and *Staphylococcus aureus* (7). Genomic comparisons allow identification of chromosomal rearrangements, including insertions, deletions, and inversions. The presence of genomic islands that may contain sequence derived from an external source such as a phage or plasmid is often detected during genomic comparisons. Since many of these genomic islands contain virulence genes (pathogenicity islands), identification of these chromosomal regions has important implications for understanding pathogenicity and virulence. Precomputed genomic comparisons of closely related strains are provided with the gMap (Genomic Map) tool (Figure 4). Syntenic blocks are detected through analysis of BLAST hits between every pair of the input sequences. Hits are split or combined to keep the number and lengths of syntenic blocks in accordance with the length of selected genomic intervals, as well as to ensure consistency of the blocks across multiple sequences. The results are displayed in a simple graphic that shows color-coded and numbered segments indicating similarity between two or more genomes. This tool can be used to visually detect chromosomal similarities, rearrangements, and the above-mentioned genomic islands, as well as smaller insertions or deletions. Care must be taken when interpreting results from incomplete genomes as hit coverage may be affected by the number of contigs and the gaps between them. Other tools that are useful for examining pairwise genomic comparisons are GenePlot and HitPlot which are linked on this page.

Protein Neighbors

To assist users in the process of discovery, and due to the time-consuming real-time searches that can be done with BLAST, all proteins are searched against the non-redundant protein database using BLAST 2.0 with the default search parameters and the results are stored in a separate database that facilitates rapid retrieval of results. A number of tools interface with the information in this database to provide specific subsets of data or to show unique relationships that are useful in the analyses of microbial genomes. These include those based on taxonomic information, chromosomal position, and similarity to proteins that belong to the COGs. Each tool is discussed in detail below with specific examples shown in Figures 4-8.

A.

B.

C.

D.

UDP-MurNAc-L-Ala

↓ MurD

UDP-MurNAc-L-Ala-D-Glu

↓ MurE

UDP-MurNAc-L-Ala-D-Glu-
meso-2,6-diaminopimelate

↓ MraY

UMP + undecaprenyl-PP-
MurNAc-L-Ala-D-Glu-meso-
2,6-diaminopimeloyl-D-Ala-D-Ala

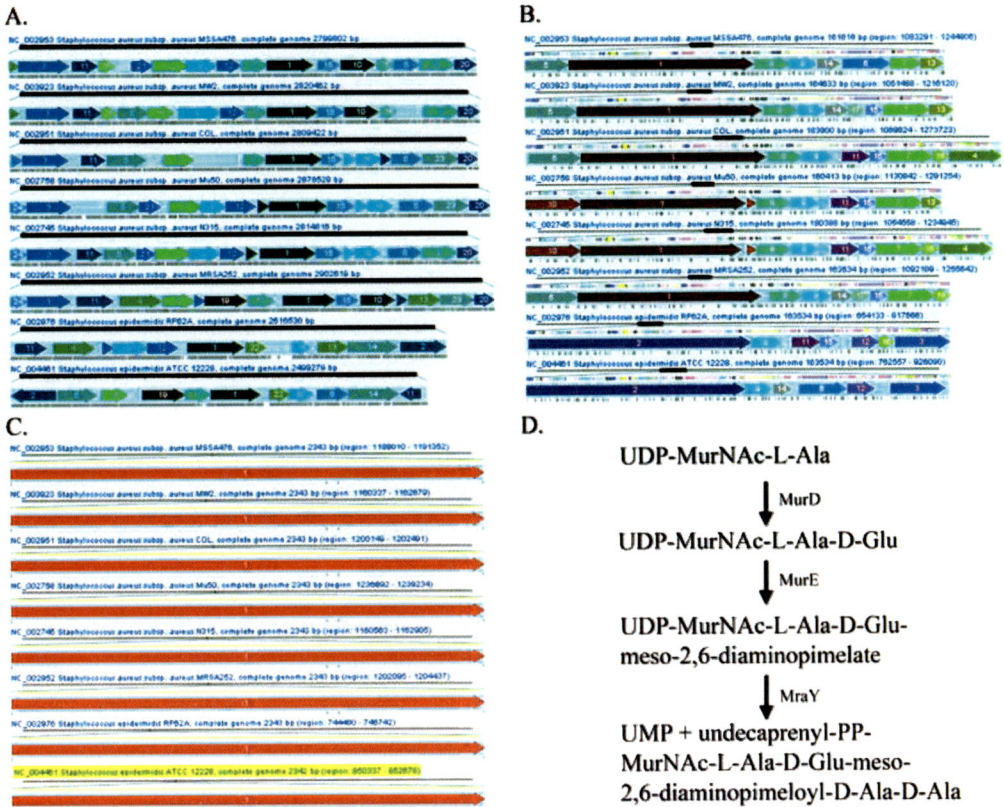

Figure 4. Genomic comparison tool (gMap) for sequence similarity. Genomic sequence are compared using BLAST and the resultant hits are filtered out to find the largest syntenic regions. Similar regions are shown color-coded and numbered in each genome with an arrow denoting the 5' to 3' direction of the hit with respect to similar segments in other genomes. Additional sequences can be added to the display by entering the Accession Number. Clicking on a given segment will magnify that region of similarity in every genome, or segments under the colored arrows can also be clicked to magnify that region from a specific organism. At that point all the segments are recalculated, recolored, and renumbered. At higher zoom levels, individual genes are drawn as rectangular bars, color-coded according to COG functional categories, and can be used to access the Entrez Protein view. Pairs of genomic sequences can be selected for output to BLAST, GenePlot, or HitPlot and any number of sequences can be removed from the list. HitPlot shows a dotplot of the two genomes selected based on the magnification level. Precomputed results are available for two categories, one for genomes from the same genera, and one for genomes based on the coverage of BLAST hits. Genomes from two or more species from the same genus may not display high levels of synteny, but similar segments in their two genomes can be found at different levels of hit coverage. An example would be the *Mycoplasma* genomes. The converse is that organisms from different genera have large syntenic blocks in their genomes such as is found in *Escherichia, Salmonella,* and *Shigella* which are all members of the *Enterobacteriaceae* family (Darling *et al.*, 2004). Genomes in both categories are grouped together based on single linkage clustering of coverage level. For example, if genome A has 75% coverage to genome B, and genome B has 75% coverage to genome C then they will all be included in a cluster at the 75% level even though the coverage between A and C may not reach the 75% level. A. gMap results for 8 *Staphylococcus* genomes including six *S. aureus* and two *S. epidermidis* genomes. B. A large syntenic region was magnified twice. The section of the genome representing this entire segment is shown as a black bar and the sequence span is indicated following the segment size. The genomic sections represented by the black bar at this magnification are approximately 150-200 Kbp in each

genome. C. A section of the *S. epidermidis* ATCC 12228 chromosome was chosen to be magnified and is highlighted. The result is an approximately 2.4 Kbp region that is similar in all *Staphylococcus* genomes. Examination of this section of the genome in all 8 species revealed that this segment contains two genes, *mraY* and *murD*, that encode UDP-MurNAc-pentapeptide phosphotransferase and UDP-N-acetylmuramoylalanine--D-glutamate ligase, respectively. These two enzymes function in cell wall metabolism and both are targets of antimicrobial agents (Silver, 2003). D. Part of the peptidoglycan biosynthetic pathway containing MurD and MraY.

A.

Nanoarchaeum equitans Kin4-M sequence   Microbial genomes ▲
genus: *Nanoarchaeum*
group: *Nanoarchaeota*
kingdom: *Archaea*

536 proteins: distribution by COGs functional categories
420 proteins can be found in COGs data base

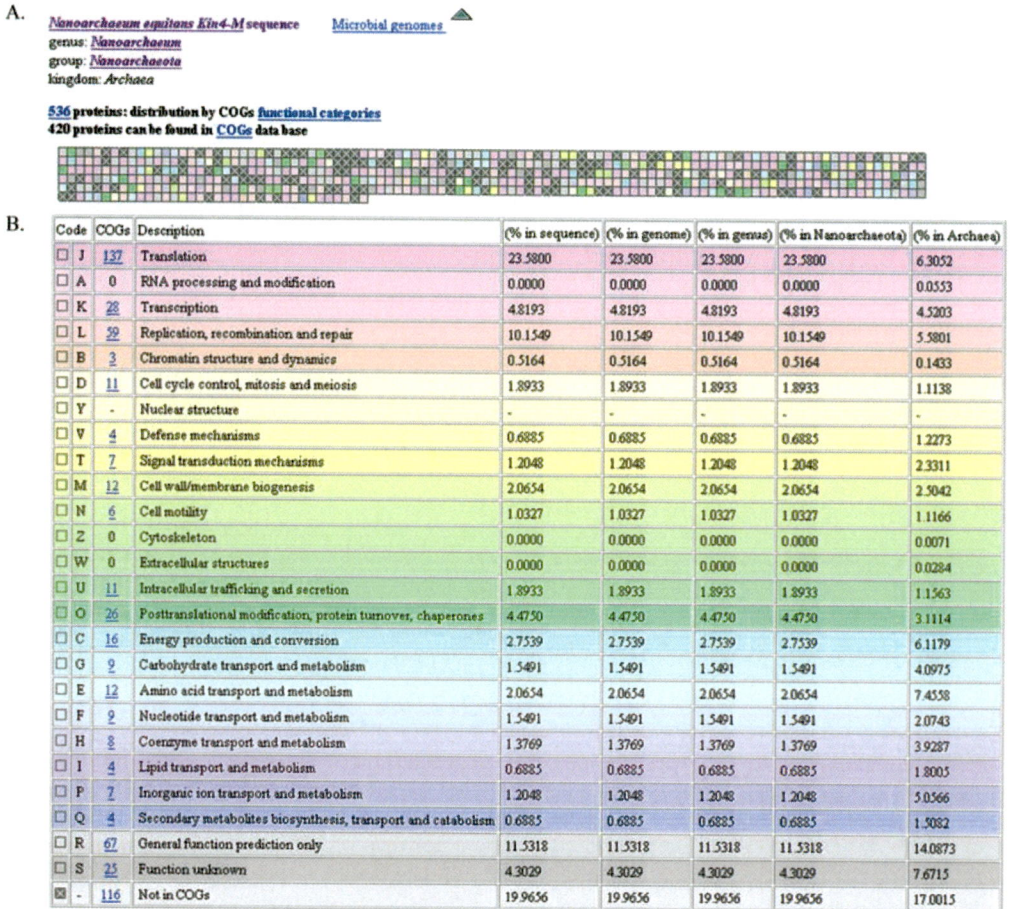

B.

Code	COGs	Description	(% in sequence)	(% in genome)	(% in genus)	(% in Nanoarchaeota)	(% in Archaea)
□ J	137	Translation	23.5800	23.5800	23.5800	23.5800	6.3052
□ A	0	RNA processing and modification	0.0000	0.0000	0.0000	0.0000	0.0553
□ K	28	Transcription	4.8193	4.8193	4.8193	4.8193	4.5203
□ L	59	Replication, recombination and repair	10.1549	10.1549	10.1549	10.1549	5.5801
□ B	3	Chromatin structure and dynamics	0.5164	0.5164	0.5164	0.5164	0.1433
□ D	11	Cell cycle control, mitosis and meiosis	1.8933	1.8933	1.8933	1.8933	1.1138
□ Y	-	Nuclear structure	-	-	-	-	-
□ V	4	Defense mechanisms	0.6885	0.6885	0.6885	0.6885	1.2273
□ T	7	Signal transduction mechanisms	1.2048	1.2048	1.2048	1.2048	2.3311
□ M	12	Cell wall/membrane biogenesis	2.0654	2.0654	2.0654	2.0654	2.5042
□ N	6	Cell motility	1.0327	1.0327	1.0327	1.0327	1.1166
□ Z	0	Cytoskeleton	0.0000	0.0000	0.0000	0.0000	0.0071
□ W	0	Extracellular structures	0.0000	0.0000	0.0000	0.0000	0.0284
□ U	11	Intracellular trafficking and secretion	1.8933	1.8933	1.8933	1.8933	1.1563
□ O	26	Posttranslational modification, protein turnover, chaperones	4.4750	4.4750	4.4750	4.4750	3.1114
□ C	16	Energy production and conversion	2.7539	2.7539	2.7539	2.7539	6.1179
□ G	9	Carbohydrate transport and metabolism	1.5491	1.5491	1.5491	1.5491	4.0975
□ E	12	Amino acid transport and metabolism	2.0654	2.0654	2.0654	2.0654	7.4558
□ F	9	Nucleotide transport and metabolism	1.5491	1.5491	1.5491	1.5491	2.0743
□ H	8	Coenzyme transport and metabolism	1.3769	1.3769	1.3769	1.3769	3.9287
□ I	4	Lipid transport and metabolism	0.6885	0.6885	0.6885	0.6885	1.8005
□ P	7	Inorganic ion transport and metabolism	1.2048	1.2048	1.2048	1.2048	5.0566
□ Q	4	Secondary metabolites biosynthesis, transport and catabolism	0.6885	0.6885	0.6885	0.6885	1.5082
□ R	67	General function prediction only	11.5318	11.5318	11.5318	11.5318	14.0873
□ S	25	Function unknown	4.3029	4.3029	4.3029	4.3029	7.6715
⊠ -	116	Not in COGs	19.9656	19.9656	19.9656	19.9656	17.0015

Figure 5. COGTable for *Nanoarchaeum equitans* Kin4-M. The entire proteome of all complete microbial genomes are searched against the proteins encoded by the 66 complete microbial genomes in COGs. Proteins encoded by genomes that are not in this group of 66 are assigned to COGs by similarity. Proteins that are above a certain cut-off score are assigned to a specific COG and as a result are also assigned to the functional category that that particular COG is associated with. A. The proteome of *Nanoarchaeum equitans* Kin4-M consists of 536 proteins, of which 420 proteins have been assigned to a COG (Waters *et al.*, 2003). The proteins are color-coded based on the COG functional category and are plotted in correlation with the gene order along the chromosome. B. All proteins encoded by *Nanoarchaeum equitans* Kin4-M are listed here, with the total number of proteins per functional category, the color assignment, and functional category. A list of proteins from this genome in each category can be obtained by clicking on the number in the COGs column. The percentage each of these represents with respect to the sequence, genome (in case there are other replicons), genus, family, and kingdom are shown as well. This allows examination of functional categories that are

under- or overrepresented in each slice. In this case, since *N. equitans* is the only complete genome from *Nanoarchaeota*, the percentages do not change until compared to the rest of the Archaea. The percentage of translation-related genes in *N. equitans* as compared to other Archaea is a reflection of the small size of the genome from this organism.

A. **7325 *Rhodopirellula baltica SH 1* proteins: taxonomic distribution of the homologs 3738 proteins can be found in CObs data base**

B.

C.

		Virus		Eukaryota		Eubacteria		Archaea			
100		7(0) V		43(1) E		61(52) B		41(4) A			
Gene		gi	score	gi	score	gi	score	gi	score	3-D	protein name
32473383 B	-			34900460	(580)	41324714	(1616)	88602629	(722)	4	cadmium-transporting ATPase
32473384 .	-			-		-		-		-	hypothetical protein RB4892
32473385 B	-			58865274	(1149)	71545741	(1820)	19915918	(1235)	40	glycogen operon protein glgX-2
32473386 .	-			-		-		-		-	hypothetical protein RB4897
32473387 .	-			-		-		-		-	hypothetical protein RB4900
32473388 .	-			-		-		-		-	hypothetical protein-transmembrane prediction
32473389 .	-			-		-		-		-	hypothetical protein RB4902
32473390 B	-			83770708	(2487)	61213822	(4276)	14324883	(105)	-	putative phosphoketolase
32473391 B	-			71013081	(346)	86356539	(564)	76801317	(130)	9	putative mutase
32473392 ⊙	-			71842251	(1231)	89899957	(1568)	2605819	(1571)	12	ATP synthase subunit B
32473393 B	-			22797529	(109)	77546663	(288)	72397719	(220)	2	ATP synthase subunit E
32473394 B	-			-		85860966	(297)	19916392	(278)	-	ATP synthase gene 1
32473395 B	-			-		69949283	(130)	19916391	(112)	-	hypothetical protein RB4910
32473396 B	-			5880710	(344)	85860964	(710)	19916390	(687)	1	ATP synthase subunit A
32473397 B	-			429174	(187)	89899962	(338)	19916389	(318)	-	ATP synthase c subunit
32473398 A	-			61393656	(144)	89899963	(374)	19916388	(413)	-	ATP synthase b subunit
32473399 B	-			1703661	(1163)	85860961	(1534)	72397713	(1458)	12	ATP synthase subunit A
32473400 A	-			24651129	(171)	89899965	(543)	19916386	(579)	3	ATP synthase gamma subunit C-terminus homolog

Figure 6. Analysis of *Rhodopirellula baltica* SH 1 using TaxTable. The entire proteome of the marine heterotroph *Rhodopirellula baltica* SH 1, an organism that is abundant in terrestrial and marine habitats, has been compared to the non-redundant database (excluding closely related species) by BLAST similarity (Glockner *et al.*, 2003). This organism is a member of the *Planctomycetales*, a monophyletic phylum that has been suggested to be one of the deepest branching phyla in Bacteria. Many of the proteins encoded by this genome have top hits to archaeal (9%) and eukaryotic (8%) proteins (Brochier and Philippe, 2002). A. One section of the genome of *Rhodopirellula baltica* SH 1 is shown, with the total number of proteins encoded by the genome (7325) and the number that have been assigned to a COG (3744) using COGnitor (see Figure 3). Each protein encoded by the genome is plotted based on chromosomal location, and color-coded based on the top BLAST hit or drawn as a cross to indicate no similarity (blue - bacteria; yellow - archaea; purple - eukaryota; gray - viruses including bacteriophages; dark green - equal hits in two or more of the other categories). B. One section of the *Rhodopirellula baltica* SH 1 genome is highlighted showing a set of genes that encode proteins with top hits to archaeal proteins. C. Gene list of the region highlighted in panel B showing protein ID, the BLAST scores to the top hits in each of the three kingdoms and viruses, and the protein name. Links to precalculated results for similarity to proteins in the structure database are also provided. There are two ATP synthase operons in this organism. The first one, ATP synthase I, is similar to other bacteria and is the gene organization is comparable. The second operon, highlighted in B and C, ATP synthase II, is more similar to proteins found in *Methanosarcina* spp

A. Pairwise genome comparison of protein homologs (symmetrical best hits)

Select two organisms to compare

| H. pylori 26695 | ▼ | versus | H. pylori J99 | ▼ |

1491 proteins total

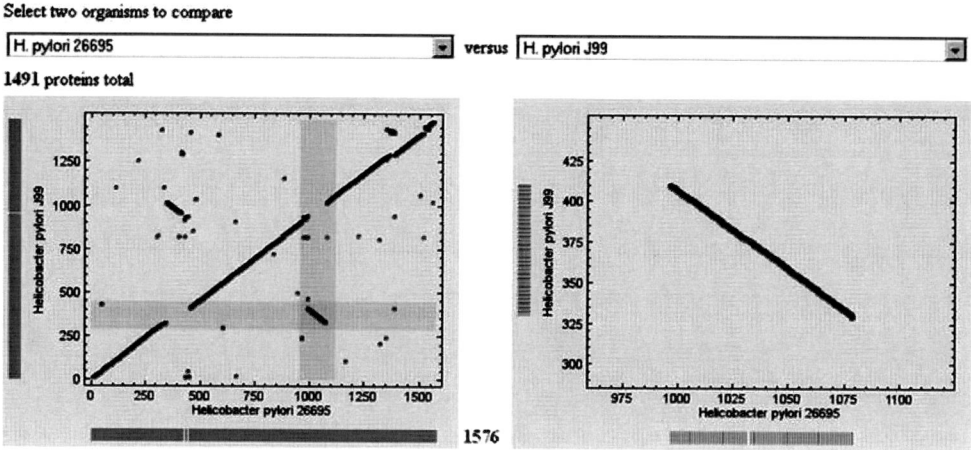

1576

B. Total number of bets 1476. Save all bets in order on genome.

bl2seq	Locus tags	Protein name
•	hp0413 - HP1010	polyphosphate kinase [Helicobacter pylori 26695]
•	hp0412 - HP1011	dihydroorotate dehydrogenase [Helicobacter pylori 26695]
•	hp0411 - HP1012	protease (pqqE) [Helicobacter pylori 26695]
•	hp0410 - HP1013	dihydrodipicolinate synthase [Helicobacter pylori 26695]
•	hp0409 - HP1014	7-alpha-hydroxysteroid dehydrogenase [Helicobacter pylori 26695]
•	hp0408 - HP1015	hypothetical protein HP1015 [Helicobacter pylori 26695]
•	hp0407 - HP1016	phosphatidylglycerophosphate synthase (pgsA) [Helicobacter pylori 26695]
•	hp0406 - HP1017	amino acid permease (rocE) [Helicobacter pylori 26695]
•	hp0405 - HP1018	hypothetical protein HP1018 [Helicobacter pylori 26695]
•	hp0405 - HP1019	serine protease (htrA) [Helicobacter pylori 26695]
•	hp0404 - HP1020	bifunctional 2-C-methyl-D-erythritol 4-phosphate cytidylyltransferase/2-C-methyl-D-erythritol 2,4-cyclodiphosphate synthase protein [Helicobacter pylori 26695]
•	hp0403 - HP1021	response regulator [Helicobacter pylori 26695]
•	hp0402 - HP1022	hypothetical protein HP1022 [Helicobacter pylori 26695]
•	hp0401 - HP1023	hypothetical protein HP1023 [Helicobacter pylori 26695]
•	hp0400 - HP1024	co-chaperone-curved DNA binding protein A (CbpA) [Helicobacter pylori 26695]
•	hp0399 - HP1025	putative heat shock protein (hspR) [Helicobacter pylori 26695]
•	hp0398 - HP1026	conserved hypothetical helicase-like protein [Helicobacter pylori 26695]
•	hp0397 - HP1027	ferric uptake regulation protein (fur) [Helicobacter pylori 26695]
•	hp0396 - HP1028	hypothetical protein HP1028 [Helicobacter pylori 26695]
•	hp0395 - HP1029	hypothetical protein HP1029 [Helicobacter pylori 26695]
•	hp0394 - HP1030	flagellar motor switch protein [Helicobacter pylori 26695]
•	hp0393 - HP1031	flagellar motor switch protein [Helicobacter pylori 26695]
•	hp0392 - HP1032	flagellar biosynthesis sigma factor FliA [Helicobacter pylori 26695]
•	hp0391 - HP1033	hypothetical protein HP1033 [Helicobacter pylori 26695]
•	hp0390 - HP1034	ATP-binding protein (ylxH) [Helicobacter pylori 26695]
•	hp0389 - HP1035	flagellar biosynthesis protein [Helicobacter pylori 26695]
•	hp0388 - HP1036	7, 8-dihydro-6-hydroxymethylpterin-pyrophosphokinase (folK) [Helicobacter pylori 26695]
•	hp0387 - HP1037	hypothetical protein HP1037 [Helicobacter pylori 26695]
•	hp0386 - HP1038	3-dehydroquinate dehydratase [Helicobacter pylori 26695]
•	hp0385 - HP1039	hypothetical protein HP1039 [Helicobacter pylori 26695]
•	hp0384 - HP1040	ribosomal protein S15 (rps15) [Helicobacter pylori 26695]
•	hp0383 - HP1041	flagellar biosynthesis protein [Helicobacter pylori 26695]
•	hp0382 - HP1042	hypothetical protein HP1042 [Helicobacter pylori 26695]
•	hp0381 - HP1043	response regulator [Helicobacter pylori 26695]
•	hp0380 - HP1044	hypothetical protein HP1044 [Helicobacter pylori 26695]

Figure 7. Analysis of two *Helicobacter pylori* strains, 26695 and J99, using Geneplot. This tool uses precalculated BLAST results to determine the top hit in two organisms. The best hit of each protein from two genomes is calculated and plotted based on chromosomal location. Similar proteins in two genomes in similar chromosomal locations will have the hit plotted along the diagonal whereas if they appear in different chromosomal locations with respect to each other, the hit will be drawn off of the diagonal. This tool can highlight insertions and deletions and large chromosomal rearrangements in closely related organisms, as well as detect syntenic regions in distantly related organisms. It is known that bacterial chromosomes can symmetrically invert around the origin of replication or the terminus and genomic comparisons of closely related species shows that this can occur often (Eisen *et al.*, 2000; Mackiewidz *et al.*, 2001). There is a large 83 Kb inversion in the *H. pylori* 26695 genome as compared to the *H. pylori* J99 genome; the continuous sequence in J99 is split into two segments in 26695 separated by approximately 600 Kb (Tomb *et al.*, 1997; Alm *et al.*, 1999). This region has been termed

the plasticity zone, as a large number of genes in this region are strain-specific and it also contains a lower GC content (35%) than the rest of the genome (39%). A. Examination of the two *H. pylori* strains using GenePlot. One of the plasticity zones is highlighted by the gray cross. The left panel shows the complete genome of both strains, while the highlighted region in the left panel is magnified in the right panel. B. Gene list of the highlighted region. Links to the bl2seq program and BLink results are available (Tatusova and Madden, 1999). Note that the locus_tags in strain 26695 are increasing numerically in this region, suggesting increasing gene order along the chromosome, while the locus_tags in strain J99 are decreasing, suggesting decreasing gene order, which agrees with the results obtained graphically.

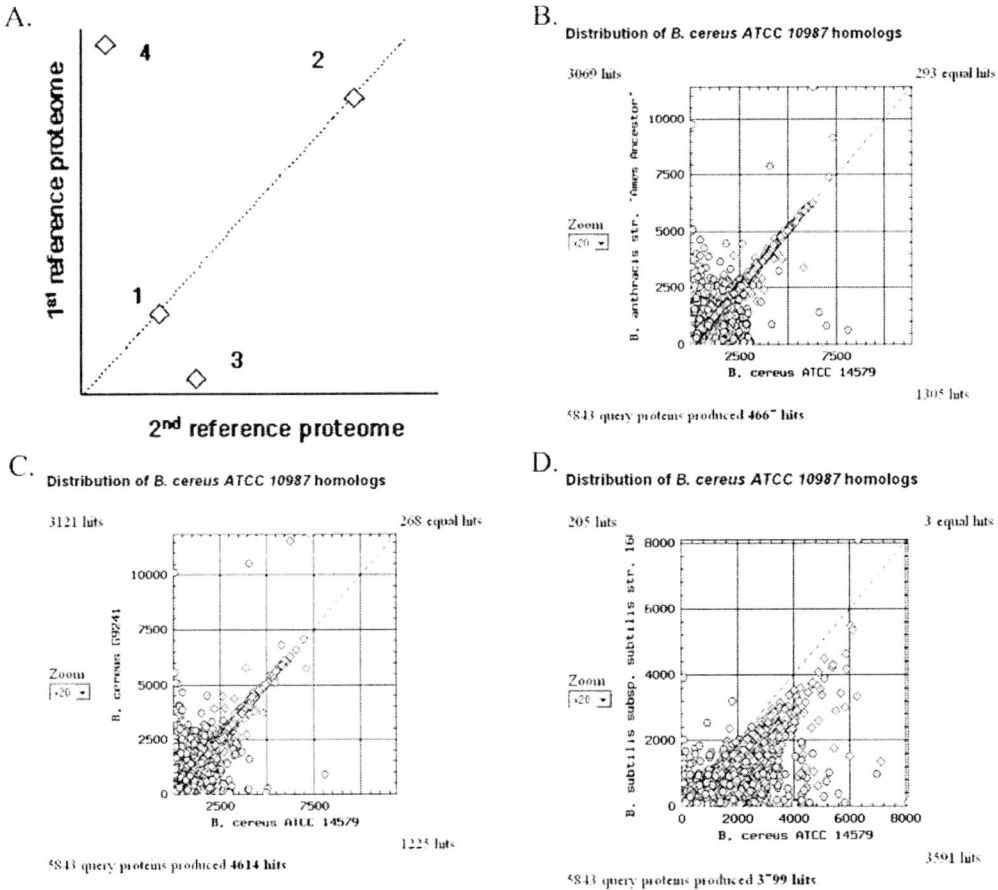

Figure 8. Three-way analysis of the proteomes of *Bacillus cereus* ATCC 10987, *B. cereus* ATCC 14579, and *B. anthracis* str. 'Ames Ancestor' using TaxPlot. This tool shows a three-way comparison of three different proteomes, including proteins encoded on all chromosomes and plasmids. Precalculated BLAST similarity results are used to plot the similarity of proteins in the query proteome based on the BLAST score to proteins in the first reference proteome (y-axis) or the second reference proteome (x-axis). A. Schematic diagram of the TaxPlot display. Proteins from the query proteome that are identically similar to proteins in the reference proteomes with low score (1) or high score (2) are drawn along the diagonal. Proteins that are asymmetrically similar are not on the diagonal: (3) low scoring protein more similar to a protein from 2nd reference proteome than a protein from 1st, (4) high scoring protein more similar to a protein from the 1st reference proteome than a protein from the 2nd. *B. cereus* and *B. anthracis* are exceptionally similar, and comparison of the genome sequence of strain ATCC 10987 to *B. anthracis* showed this (Ivanova *et al.*, 2003; Rasko *et al.*, 2004). B. The TaxPlot of *B.*

cereus 10987 (query), *B. anthracis* str. 'Ames Ancestor' (reference 1), and *B. cereus* ATCC 14579 (reference 2), confirms this and is similar to what is observed when *B. anthracis* is replaced with *B. cereus* G2941 in C. Using a more distantly related species, *B. subtilis* subsp. *subtilis* str. 168, shows that many of the proteins in strain ATCC 10987 are more similar to proteins from strain ATCC 14579, as expected and shown in D. Further analysis of *B. anthracis* G2941 showed that it causes a disease similar to anthrax and harbors a plasmid that contains similar toxin genes to the *B. anthracis* plasmid pXO1 while strain ATCC 10987 has a plasmid similar to pXO1 that does not encode toxin proteins (Hoffmaster *et al.*, 2004). The major differences between *B. cereus* and *B. anthracis* that affect virulence are due to plasmid differences and not chromosomal alterations.

BLink

BLink (BLAST Link) provides access to a graphically rich set of tools for analyzing all precalculated protein neighbors. Closely related proteins are easily identifiable as are both the BLAST score and the position and length of each hit. The organism that encodes the protein is shown as belonging to seven large taxonomic groups consisting of Archaea, Bacteria, Metazoa, Fungi, Plants, Viruses, and other Eukaryotes. The hits are filterable so that these major taxonomic groups can be easily added or removed from the hit list and the list of hits are also sortable by taxonomic proximity which means that proteins from related species can be found easily. Furthermore, sequences are filterable based on the source database, and hits from the Protein Data Bank (PDB), SwissProt, RefSeq, complete genomes, or COG members can be selected. Similarity to known structures and domain searches against the Conserved Domain Database can also be initiated. BLink is found on the protein record and through Gene reports.

Cluster of Orthologous Groups

The clusters of orthologous groups attempts to identify orthologous genes and to visualize protein clusters in both prokaryotes and eukaryotes (Tatusov *et al.*, 2001). The current version of the database (COG) includes orthologous groups from 66 completely sequenced organisms, and a eukaryotic version, KOG, includes 7 eukaryotes (Tatusov *et al.*, 2003). COGs are constructed from an all-against-all sequence similarity comparison using BLAST, resulting in the construction of orthologous groups that consist of proteins from at least three phylogenetic lineages that have been broken down into 18 functional groups (Tatusov *et al.*, 1997). COG alignments are now included in the Conserved Domain Database.

The COGTable provides a useful look at the entire genome of a given organism broken down by COG functional categories (Figure 5). The display consists of two parts. The first part displays the proteins color-coded with respect to COG functional category drawn on a map that represents the order of genes along the chromosome which is useful in finding functionally related genes in specific chromosomal locations. The second part of the display is a list of the COG functional categories and the number of proteins that are found in each category.

Taxtable

The TaxTable provides a taxonomic distribution of precalculated BLAST protein similarity (Tatusova *et al.*, 1999). Proteins from closely related species are excluded, which allows a broad taxonomic view of proteome similarity. Proteins are color-coded according to their best hit to proteins from the three kingdoms and viruses; bacteria (blue), archaea (yellow), eukarya (red), and viruses and bacteriophages (gray) (Figure 6). Species-specific proteins or proteins that do not have similarity to the proteins in the nr BLAST database are not colored. A configurable cut-off score allows inclusion or exclusion of more distantly related proteins and is available through both a numerical value or a graph based alignment that shows the distribution of hits to proteins from the three kingdoms as well as viruses. The best hits from each of the taxonomic groups are available as a downloadable table from the graphical display. This table provides a quick visual take on the similarity of the proteome by broad taxonomic group and can also be used for the identification of potentially horizontally transferred genes or in finding genomic islands.

Geneplot

Geneplot combines protein-sequence similarity searches with sequence location, unlike gMap, which is solely based on nucleotide sequence similarity. This tool can be used to detect syntenic regions or chromosomal rearrangements in closely related species, as well as contiguous regions in distantly related organisms (Figure 7). Small insertions or deletions and major genomic islands in closely related species are also identifiable using this tool and a table of the best hits between both organisms is available. Geneplot provides a more detailed view of the pairwise comparison of two genomes.

Taxplot

TaxPlot compares two reference proteomes to a query proteome and thus provides a three-way comparison of proteome similarities (Figure 8). A single protein can be searched for and highlighted, and entire COG functional categories can be examined, allowing a three-way comparison of a single category of proteins. This tool can be used to detect potentially horizontally transferred genes by using a distantly related organism in comparison to two closely related strains (Figure 9).

Genomes to Domains

Domains are discrete units that form distinct folding regions in the final protein tertiary structure and perform unique functions. An analysis of the domain structure of proteins can provide insight into function. The Conserved Domain Database contains a set of PSI-BLAST-derived position-specific scoring matrices that represent domains taken from the

Simple Modular Architecture Research Tool (Smart), Pfam, and from domain alignments derived from COGs (Marchler-Bauer *et al.*, 2005). Dynamic queries are performed using the CD-Search function, which is also done each time a BLAST search is initiated. The domain architecture of a specific protein can be examined using the Conserved Domain Architecture Retrieval Tool (CDART) which graphically displays the various domains that are found in the query. Existing 3D structures for a given domain can be viewed using the 3D molecular structure viewer, Cn3D, and alignments can be built using this tool (Wang *et al.*, 2000B). Pre-computed domain similarity searches are available for complete microbial genomes in the form of the CDD table which is organized either by domain or by position of the coding region along the chromosome.

A.

Distribution of *M. leprae* TN homologs

B.

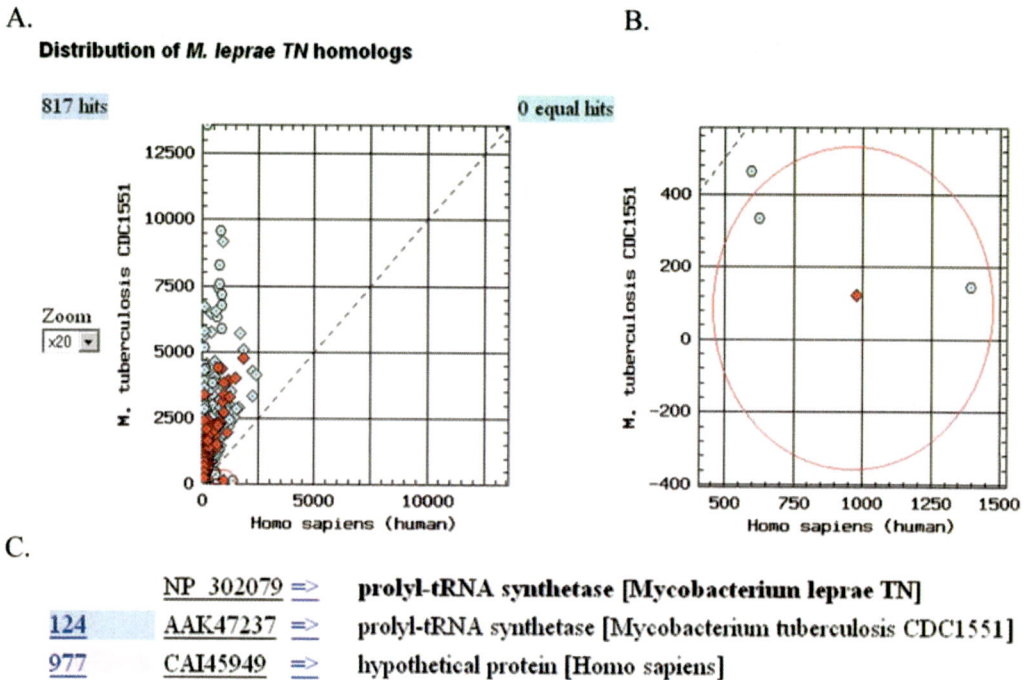

C.

Figure 9. Comparison of the proteomes of two *Mycobacterium* strains and *Homo sapiens* using TaxPlot. This tool can also be used to determine potential horizontally transferred genes. Using two similar organisms, one as the reference proteome and the other as the query proteome, and a phylogenetically distant species as the second reference proteome can highlight lateral gene transfer events. It was suspected that the *proS* gene of *Mycobacterium leprae* TN that encodes the prolyl-tRNA synthetase is an example of horizontal transfer from host to pathogen. ProS is more similar to eukaryotic proteins than proteins encoded by other *Mycobacterium* genomes (Wolf *et al.*, 1999; Cole *et al.*, 2001). A. TaxPlot of *M. leprae* TN, *M. tuberculosis* CDC1551, and human (by using the taxid, 9606, that corresponds to *Homo sapiens*) demonstrates this. Proteins that fall into the translationally-related functional category were highlighted. One protein is more similar to a human protein than to a protein found in a related species. B. The outlier in A is highlighted, and the similarity scores to proteins in both *M. tuberculosis* and human is shown in C. *M. leprae* may have obtained the gene from its host, since *M. leprae* infects humans and causes leprosy. The BLAST score of ProS is much higher to the human protein than to the *M. tuberculosis* ortholog, suggesting a horizontal transfer even during the evolutionary history of *M. leprae*.

Genomes to Structures

As with the domains noted above, it is useful to know that a given protein is similar in sequence to a known structure. The structural database at NCBI, MMDB (Molecular Modeling Database), contains three-dimensional structures processed from the Protein Data Bank (PDB) and analysis tools for visualization and comparative analysis (Deshpande *et al.*, 2005; Marchler-Bauer *et al.*, 2005; Wang *et al.*, 2000A; Wheeler *et al.*, 2005). With the advent of high-throughput structural genomics for a number of commonly studied microbes such as *Escherichia coli*, *Mycobacterium tuberculosis*, and *Haemophilus influenzae*, the number of structures is increasing rapidly (Schmid, 2004). Similarities to known structures are available in the PDBTable. The webview provides a configurable cut-off score that allows users to include or exclude more distantly related structures. Sequence and structural similarity for a given protein can be further examined by doing domain searching and structural similarity using VAST (Vector Alignment Search Tool (Madej *et al.*, 1995; Gibrat *et al.*, 1996).

Conclusion

The tremendous increase in genomic data in the last ten years has greatly expanded our understanding of biology. Microbial sequencing projects now span from incomplete genomes, complete genomes, large-scale comparative genomic projects of multiple strains, and to the new field of metagenomics where the entire complement of DNA from a given ecological niche is being sequenced. Although these provide an ever greater resource for studying biology, there is still a lag from the initial submission of sequence data to biological understanding. In an effort to promote biological annotation and analyses NCBI provides a number of tools for genomic sequence comparison as well as protein sequence, domain, and structural similarity searches. The results are available from single entries in Entrez to interactive webviews that allow configuration of the results. Output from these tools can be downloaded in bulk through the web or via FTP, and the E-Utilities function allows users the ability to build highly specific and unique pipelines for the analysis of data. All of these tools are an attempt to provide users with additional layers of information than what was found on original data submissions, thus enabling researchers to find the most up-to-date and relevant information available. NCBI continues to add new databases, information, and tools to address the issue of ever increasing amounts of data.

References

Akman, L., Yamashita, A., Watanabe, H., *et al.* (2002). Genome sequence of the endocellular obligate symbiont of tsetse flies, Wigglesworthia glossinidia. *Nat Genet.* 32(3): 402-7.

Alfarano, C., Andrade, C. E., Anthony, K., *et al.* (2005). The Biomolecular Interaction Network Database and related tools 2005 update. *Nucleic Acids Res.* 33 Database Issue: D418-24.

Alm, R. A., Ling, L. S., Moir, D. T., *et al.* (1999). Genomic-sequence comparison of two unrelated isolates of the human gastric pathogen Helicobacter pylori. *Nature.* 397(6715): 176-80.

Altschul, S. F., Gish, W., Miller, W., *et al.* (1990). Basic local alignment search tool. *J Mol Biol.* 215(3): 403-10.

Altschul, S. F., Madden, T. L., Schaffer, A. A., *et al.* (1997). Gapped BLAST and PSI-BLAST: a new generation of protein database search programs. *Nucleic Acids Res.* 25(17): 3389-402.

Bairoch, A. (2000). The ENZYME database in 2000. *Nucleic Acids Res.* 28(1): 304-5.

Barrett, T., Suzek, T. O., Troup, D. B., *et al.* (2005). NCBI GEO: mining millions of expression profiles--database and tools. *Nucleic Acids Res.* 33 Database Issue: D562-6.

Benson, D. A., Karsch-Mizrachi, I., Lipman, D. J., *et al.* (2005). GenBank. *Nucleic Acids Res.* 33 Database Issue: D34-8.

Bentley, S. D., Chater, K. F., Cerdeno-Tarraga, A. M., *et al.* (2002). Complete genome sequence of the model actinomycete Streptomyces coelicolor A3(2). *Nature.* 417(6885): 141-7.

Bernal, A., Ear, U. and Kyrpides, N. (2001). Genomes OnLine Database (GOLD): a monitor of genome projects world-wide. *Nucleic Acids Res.* 29(1): 126-7.

Besemer, J., Lomsadze, A. and Borodovsky, M. (2001). GeneMarkS: a self-training method for prediction of gene starts in microbial genomes. Implications for finding sequence motifs in regulatory regions. *Nucleic Acids Res.* 29(12): 2607-18.

Brochier, C. and Philippe, H. (2002). Phylogeny: a non-hyperthermophilic ancestor for bacteria. *Nature.* 417(6886): 244.

Cole, S. T., Eiglmeier, K., Parkhill, J., *et al.* (2001). Massive gene decay in the leprosy bacillus. *Nature.* 409(6823): 1007-11.

Cummings, L., Riley, L., Black, L., *et al.* (2002). Genomic BLAST: custom-defined virtual databases for complete and unfinished genomes. *FEMS Microbiol Lett.* 216(2): 133-8.

Curtis, T. P., Sloan, W. T. and Scannell, J. W. (2002). Estimating prokaryotic diversity and its limits. *Proc Natl Acad Sci U S A.* 99(16): 10494-9.

Darling, A. C., Mau, B., Blattner, F. R., *et al.* (2004). Mauve: multiple alignment of conserved genomic sequence with rearrangements. *Genome Res.* 14(7): 1394-403.

Delcher, A. L., Harmon, D., Kasif, S., *et al.* (1999). Improved microbial gene identification with GLIMMER. *Nucleic Acids Res.* 27(23): 4636-41.

Deshpande, N., Addess, K. J., Bluhm, W. F., *et al.* (2005). The RCSB Protein Data Bank: a redesigned query system and relational database based on the mmCIF schema. *Nucleic Acids Res.* 33 Database Issue: D233-7.

Dykhuizen, D. E. (1998). Santa Rosalia revisited: why are there so many species of bacteria? *Antonie Van Leeuwenhoek.* 73(1): 25-33.

Eisen, J. A., Heidelberg, J. F., White, O., *et al.* (2000). Evidence for symmetric chromosomal inversions around the replication origin in bacteria. *Genome Biol.* 1(6): RESEARCH0011.

Fleischmann, R. D., Adams, M. D., White, O., *et al.* (1995). Whole-genome random sequencing and assembly of Haemophilus influenzae Rd. *Science.* 269(5223): 496-512.

Fraser, C. M., Casjens, S., Huang, W. M., *et al.* (1997). Genomic sequence of a Lyme disease spirochaete, Borrelia burgdorferi. *Nature.* 390(6660): 580-6.

Gibrat, J. F., Madej, T. and Bryant, S. H. (1996). Surprising similarities in structure comparison. *Curr Opin Struct Biol.* 6(3): 377-85.

Glockner, F. O., Kube, M., Bauer, M., *et al.* (2003). Complete genome sequence of the marine planctomycete Pirellula sp. strain 1. *Proc Natl Acad Sci U S A.* 100(14): 8298-303.

Hallam, S. J., Putnam, N., Preston, C. M., *et al.* (2004). Reverse methanogenesis: testing the hypothesis with environmental genomics. *Science.* 305(5689): 1457-62.

Handelsman, J. (2004). Metagenomics: application of genomics to uncultured microorganisms. *Microbiol Mol Biol Rev.* 68(4): 669-85.

Hoffmaster, A. R., Ravel, J., Rasko, D. A., *et al.* (2004). Identification of anthrax toxin genes in a Bacillus cereus associated with an illness resembling inhalation anthrax. *Proc Natl Acad Sci U S A.* 101(22): 8449-54.

Ikeda, H., Ishikawa, J., Hanamoto, A., *et al.* (2003). Complete genome sequence and comparative analysis of the industrial microorganism Streptomyces avermitilis. *Nat Biotechnol.* 21(5): 526-31.

Ivanova, N., Sorokin, A., Anderson, I., *et al.* (2003). Genome sequence of Bacillus cereus and comparative analysis with Bacillus anthracis. *Nature.* 423(6935): 87-91.

Kaneko, T., Nakamura, Y., Sato, S., *et al.* (2002). Complete genomic sequence of nitrogen-fixing symbiotic bacterium Bradyrhizobium japonicum USDA110. *DNA Res.* 9(6): 189-97.

Kanz, C., Aldebert, P., Althorpe, N., *et al.* (2005). The EMBL Nucleotide Sequence Database. *Nucleic Acids Res.* 33 Database Issue: D29-33.

Keseler, I. M., Collado-Vides, J., Gama-Castro, S., *et al.* (2005). EcoCyc: a comprehensive database resource for Escherichia coli. *Nucleic Acids Res.* 33 Database Issue: D334-7.

Mackiewicz, P., Mackiewicz, D., Kowalczuk, M., *et al.* (2001). Flip-flop around the origin and terminus of replication in prokaryotic genomes. *Genome Biol.* 2(12): INTERACTIONS1004.

Madej, T., Gibrat, J. F. and Bryant, S. H. (1995). Threading a database of protein cores. *Proteins.* 23(3): 356-69.

Maglott, D., Ostell, J., Pruitt, K. D., *et al.* (2005). Entrez Gene: gene-centered information at NCBI. *Nucleic Acids Res.* 33 Database Issue: D54-8.

Marchler-Bauer, A., Anderson, J. B., Cherukuri, P. F., *et al.* (2005). CDD: a Conserved Domain Database for protein classification. *Nucleic Acids Res.* 33 Database Issue: D192-6.

McGinnis, S. and Madden, T. L. (2004). BLAST: at the core of a powerful and diverse set of sequence analysis tools. *Nucleic Acids Res.* 32(Web Server issue): W20-5.

Miyazaki, S., Sugawara, H., Ikeo, K., *et al.* (2004). DDBJ in the stream of various biological data. *Nucleic Acids Res.* 32 Database issue: D31-4.

Pruitt, K. D., Tatusova, T. and Maglott, D. R. (2005). NCBI Reference Sequence (RefSeq): a curated non-redundant sequence database of genomes, transcripts and proteins. *Nucleic Acids Res.* 33 Database Issue: D501-4.

Rappe, M. S. and Giovannoni, S. J. (2003). The uncultured microbial majority. *Annu Rev Microbiol.* 57: 369-94.

Rasko, D. A., Ravel, J., Okstad, O. A., *et al.* (2004). The genome sequence of Bacillus cereus ATCC 10987 reveals metabolic adaptations and a large plasmid related to Bacillus anthracis pXO1. *Nucleic Acids Res.* 32(3): 977-88.

Read, T. D., Salzberg, S. L., Pop, M., *et al.* (2002). Comparative genome sequencing for discovery of novel polymorphisms in Bacillus anthracis. *Science.* 296(5575): 2028-33.

Salzberg, S. L., Church, D., DiCuccio, M., *et al.* (2004). The genome Assembly Archive: a new public resource. *PLoS Biol.* 2(9): E285.

Schmid, M. B. (2004). Seeing is believing: the impact of structural genomics on antimicrobial drug discovery. *Nat Rev Microbiol.* 2(9): 739-46.

Silver, L. L. (2003). Novel inhibitors of bacterial cell wall synthesis. *Curr Opin Microbiol.* 6(5): 431-8.

Tatusov, R. L., Fedorova, N. D., Jackson, J. D., *et al.* (2003). The COG database: an updated version includes eukaryotes. *BMC Bioinformatics.* 4(1): 41.

Tatusov, R. L., Koonin, E. V. and Lipman, D. J. (1997). A genomic perspective on protein families. *Science.* 278(5338): 631-7.

Tatusov, R. L., Natale, D. A., Garkavtsev, I. V., *et al.* (2001). The COG database: new developments in phylogenetic classification of proteins from complete genomes. *Nucleic Acids Res.* 29(1): 22-8.

Tatusova, T. A., Karsch-Mizrachi, I. and Ostell, J. A. (1999). Complete genomes in WWW Entrez: data representation and analysis. *Bioinformatics.* 15(7-8): 536-43.

Tatusova, T. A. and Madden, T. L. (1999). BLAST 2 Sequences, a new tool for comparing protein and nucleotide sequences. *FEMS Microbiol Lett.* 174(2): 247-50.

Tomb, J. F., White, O., Kerlavage, A. R., *et al.* (1997). The complete genome sequence of the gastric pathogen Helicobacter pylori. *Nature.* 388(6642): 539-47.

Torsvik, V., Ovreas, L. and Thingstad, T. F. (2002). Prokaryotic diversity--magnitude, dynamics, and controlling factors. *Science.* 296(5570): 1064-6.

Tyson, G. W., Chapman, J., Hugenholtz, P., *et al.* (2004). Community structure and metabolism through reconstruction of microbial genomes from the environment. *Nature.* 428(6978): 37-43.

Venter, J. C., Remington, K., Heidelberg, J. F., *et al.* (2004). Environmental genome shotgun sequencing of the Sargasso Sea. *Science.* 304(5667): 66-74.

Wang, Y., Bryant, S., Tatusov, R., *et al.* (2000A). Links from genome proteins to known 3-D structures. *Genome Res.* 10(10): 1643-7.

Wang, Y., Geer, L. Y., Chappey, C., *et al.* (2000B). Cn3D: sequence and structure views for Entrez. *Trends Biochem Sci.* 25(6): 300-2.

Waters, E., Hohn, M. J., Ahel, I., *et al.* (2003). The genome of Nanoarchaeum equitans: insights into early archaeal evolution and derived parasitism. *Proc Natl Acad Sci U S A.* 100(22): 12984-8.

Wernegreen, J. J. and Moran, N. A. (1999). Evidence for genetic drift in endosymbionts (Buchnera): analyses of protein-coding genes. *Mol Biol Evol.* 16(1): 83-97.

Wheeler, D. L., Barrett, T., Benson, D. A., *et al.* (2005). Database resources of the National C

Wolf, Y. I., Aravind, L., Grishin, N. V., *et al.* (1999). Evolution of aminoacyl-tRNA synthetases--analysis of unique domain architectures and phylogenetic trees reveals a complex history of horizontal gene transfer events. *Genome Res.* 9(8): 689-710.

In: In Silico Genomics and Proteomics
Editors: N. Mulder and R. Apweiler, pp. 185-200

ISBN 1-59454-995-8
© 2006 Nova Science Publishers, Inc.

Chapter 13

PlasmoDB: The Plasmodium Genomics and Functional Genomics Resource

Patricia L. Whetzel, Shailesh V. Date, Kobby Essien,
Martin J. Fraunholz, Bindu Gajria, Gregory R. Grant, John Iodice,
Jessica C. Kissinger, Philip T. Labo, Arthur J. Milgram,
David S. Roos and Christian J. Stoeckert Jr.,[*]

Genomics Institute, Center for Bioinformatics, and Departments of Biology and Genetics,
University of Pennsylvania, Philadelphia PA 19104 USA; Center for Tropical and
Emerging Global Diseases, and Department of Genetics, University of Georgia, Athens,
Georgia 30602 USA

Abstract

PlasmoDB (http://PlasmoDB.org) provides proteomic data along with genomic and other
functional genomic datasets for the malaria parasite, *Plasmodium falciparum*. Types of
proteomic data available include mass spectrometry of parasite life stages, known and
predicted protein structures, and a variety of sequence-based protein features. These data
can be viewed in the context of individual Gene Pages or used as the basis for complex
queries to identify target proteins. It is through the Gene Pages and queries that
PlasmoDB provides an integrated view of malaria parasite genes, their expression, their
protein products, and their interactions.

Introduction

The *Plasmodium* Genome Database (http://PlasmoDB.org) was created in 2000 to
provide the malaria research community access to genomic-scale datasets. PlasmoDB
originally focused on automated analysis of available sequence data (The Plasmodium

---

[*] To whom correspondence should be address, at : stoeckrt@pcbi.upenn.edu

Database Collaborative, 2001; Bahl et al., 2002; Milgram et al., 2003). With the release of a reference sequence for *P. falciparum* strain 3D7 (Gardner et al., 2002), the focus of PlasmoDB has shifted to providing access to completed sequence and curated annotations, compiling all available information associated with individual genes (predicted gene and protein sequence and structure, annotations, etc), and enabling genome-wide studies – such as the identification of all proteins with a particular sequence motif, or predicted genes at a chromosomal location (Kissinger et al., 2002; Bahl et al., 2003).

The availability of an effectively complete *P. falciparum* genome sequence has stimulated a wide range of functional genomics research (cf. Bozdech et al., 2003; Coppel et al., 2004; Doolan et al., 2003; Florens et al., 2002; Florens et al., 2004; Hiller et al., 2004; Lasonder et al., 2002; Le Roch et al., 2003; Le Roch et al., 2004; Marti et al., 2004; Roos et al., 2002; Sam-Yellowe et al., 2004), and PlasmoDB has endeavored to keep pace with these studies, providing access to the underlying datasets, and allowing a variety of integrative queries – e.g. finding all genes for which both transcript and proteomics data suggests expression in gametocyte stage parasites (Kissinger and Roos, 2004).

In parallel with the rapid expansion of genomic-scale functional genomics datasets for *P. falciparum,* sequencing efforts have been dedicated to several other *Plasmodium* species, including *P. berghei, chabaudi, gallinaceum, knowlesi, reichenowi, vivax,* and *yoelii* (Berry et al., 2004; Carlton et al., 2003; Carlton et al., 2005; Hall et al., 2005). Accordingly, pipelines for the automated analysis of unfinished sequences continue to be important, and PlasmoDB has grown to include genes from additional *Plasmodium* species. Functional genomics data for these organisms is also beginning to emerge (Hall et al., 2005; Kaiser et al., 2004).

PlasmoDB presents genomic data via Gene Pages, and also incorporates a variety of analysis tools and data mining queries, which take advantage of the underlying database structure. Data of interest can be saved using the History feature and downloaded using the ReportMaker tool. The descriptions in this chapter refer to features and functions present in version 4.4 of PlasmoDB and some of the details may change in future versions. PlasmoDB will continue to integrate new datasets as they emerge, developing tools of interest to the malaria researcher, and facilitating the discovery of new diagnostics, drugs and vaccines.

Gene Pages

Gene Pages provide access to the complete collection of information for each gene in the PlasmoDB database. These can be accessed individually by queries on the home page for the different species or through data mining queries discussed in the next section. Multiple alternative views of the data are provided (Figure 1). The Summary view presents the most commonly requested information about the gene in question, including curated annotations, sequence similarities to proteins in the GenBank non-redundant database (Benson et al., 2005) based on BLAST results (Basic Local Alignment Tool; Altschul et al., 1997), Ortholog/Paralog predictions (Li et al., 2003), a graphical representation of the genomic context and predicted protein features (including secondary structure predictions), and highlights of functional genomics data such as microarray and proteomics results. More detailed information about genes is obtained using the specialized views (Annotation, Protein, Expression, Sequence), and can be accessed with a single click from the Summary page.

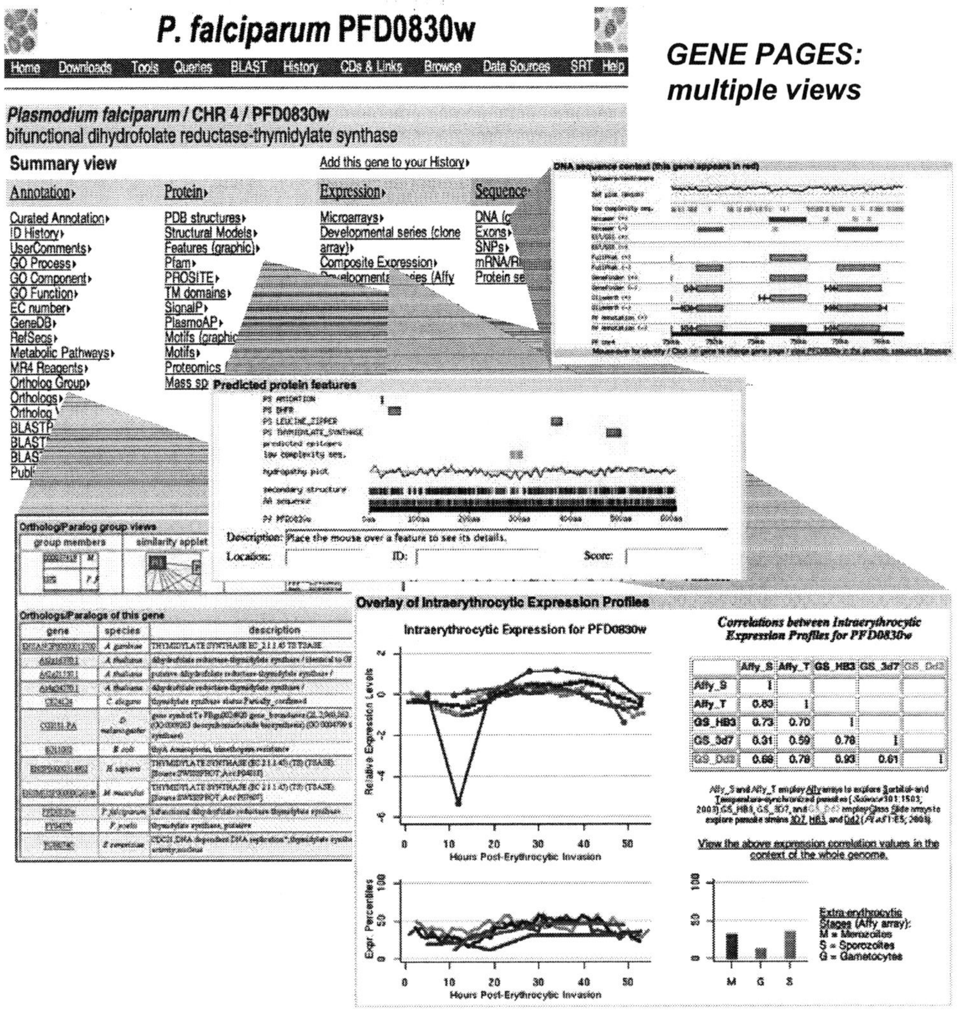

Figure 1. Gene Pages on PlasmoDB. The summary view (upper left) provides a tabulation of the various forms of information available, and scrolls to reveal highlights of general interest, such as curated annotations (and comments from the research community, which may also be entered by the user), DNA and protein features, transcript and protein expression data, information on protein structure, reagents, publications, etc. Insets illustrate, for the *P. falciparum* DHFR-TS gene (PFD0830w): (*i*) a list of orthologs in other species, (*ii*) mapped protein motifs and features, (*iii*) transcript profiling data, including information on both abundance and induction/repression, based on both photolithographic and glass slide microarrays, and (*iv*) the gene model(s), sequence, and information on local chromosomal DNA context.

The Annotation view displays all known information related to a gene's annotation, including 'official' annotations from curators at the genome sequencing centers (Berry et al., 2004), 'unofficial' comments from PlasmoDB users, and a variety of automated analyses. This view also provides the history of the gene identifier. As sequencing efforts proceed, provisional names are often used to identify genes, before being assigned permanent identifiers once the genome assemblies are completed and annotated. For example, in the name PFF0480w, 'PFF' denotes a gene on *P. falciparum* chromosome VI (PFA = chr I, PFF = chr VI, etc), 0480 reflects gene position from left to right on the chromosome, and 'w' vs. 'c' reflects the coding strand. Prior to assigning this stable identifier based on the complete genome sequence, however, PFF0480w was known as MAL6P1.100. The Annotation view also displays assignments based on structured ontologies, including enzyme commission (EC) numbers, and Gene Ontology (GO) terms describing biological process, cellular component and function (Harris et al., 2004, http://www.geneontology.org). Additional information available under the Annotation view includes RefSeq assignments, a graphical view of BLAST matches to the non-redundant GenBank/EMBL/DDBJ sequence databases for all species, and predicted orthologs and paralogs (Li et al., 2003). Links are also provided to external resources, including the GeneDB database (Hertz-Fowler et al., 2004, http://www.GeneDB.org) maintained by the Sanger Institute's Pathogen Sequencing Unit, including data for *P. falciparum, P. berghei, P. chabaudi P. knowlesi*, and many other organisms. Additional information may be obtained from links with the Malaria Parasite Metabolic Pathway database at Hebrew University (http://sites.huji.ac.il/malaria/), and the MR4 reagent repository at the American Type Culture Collection (http://www.malaria. mr4.org).

Protein views link to any structures available in the Protein Data Bank, and models for *Plasmodium* orthologs of proteins in other species for which structural information is available (http://bioinfo.icgeb.res.in/codes/model.html). This view also lists – and provides a graphical representation of – predicted protein secondary structure based on PsiPred (a neural network based algorithm using position-specific scoring matrices; McGuffin et al., 2000), predicted Pfam (Bateman et al., 2004) and ProSite (Hulo et al., 2004) domains, putative signal peptides, transmembrane domains, and other targeting predictions (Bender et al, 2003; Foth et al, 2003; Hiller et al, 2004; Marti et al, 2004), such as those mediating targeting to the apicoplast, a promising drug target (Roos et al., 2002; Ralph et al., 2004). In addition, the Protein view shows BLASTP results comparing the protein against the motif databases ProDom and CDD (Servant et al., 2002). Protein expression data is provided as a map illustrating the location of peptide fragments identified through mass spectrometry studies of different developmental stages (Florens et al., 2002) or subcellular fractions (Florens et al., 2004).

Expression views display information from both RNA microarrays (on a variety of platforms) and proteomics data. Three distinct probe sets – clone arrays (Mamoun et al., 2001), spotted oligonucleotide arrays (Bozdech et al., 2003), and photolithographic (Affymetrix) oligonucleotide arrays (Le Roch et al., 2003) – have been mapped to *P. falciparum* genome. Several array studies have been loaded into PlasmoDB and are available for analysis, including both glass slide and photolithographic arrays interrogating whole genome expression throughout the intraerythrocytic cycle, and entering into gametocytogenesis. Oligonucleotides that map to a given gene are displayed in the microarray gene page view, along with a graphical representation of expression at various life cycle stages.

Sequence views provide a graphical representation of predicted genes, %AT plot and predictions of low complexity sequence. The Sequence view also presents exon predictions (based on a variety of algorithms, along with curated annotation), information about any single nucleotide polymorphisms (SNPs) identified for this gene (limited SNP data is available at present, but this class of data is likely to become increasingly prominent in the coming years), and predicted mRNA and protein sequences. As with all results found in other gene views, more detailed information – along with analysis of the entire genome – can be gleaned by querying the data.

Analysis Tools

PlasmoDB provides the user with a variety of analysis tools for examining – and extracting information from – the genome and predicted proteome, using BLAST, electronic polymerase chain reactions (ePCR; Schuler, GD, 1997), defined motif searches, and tools for the analysis of microarray and proteomics data.

BLAST tools (Altschul et al., 1997) available at PlasmoDB allow users to search against a variety of nucleotide and amino acid databases of *P. falciparum* and additional *Plasmodium* species. The results of these searches are stored in a History page, from which they may be downloaded or combined with other queries for further analysis (see below). An oligonucleotide search tool uses the WU-BLAST algorithm, with parameters optimized for query sequences as short as 15-20 bp. A further tool provides a graphical view of sequence similarity as a function of taxonomic divergence, i.e. the similarity between a given gene and genes in other organisms of varying taxonomic relatedness.

The ePCR tool permits specified sequence-tagged site (STS) markers to be identified within finished and draft genome sequences; other tools allow the user to search for oligonucleotide locations based on the name of the sequence probe. Given a user-defined span (or spans) of genomic DNA, sequence(s) may be downloaded using the multiple sequence retrieval tool (SRT), or a list of features may be compiled for that region. PlasmoDB also includes tools to search for motifs, including organellar targeting signals or motifs defined by the user. For example, the investigator may wish to identify any predicted protein with multiple instances where two cysteines are spaced 11 amino acids apart.

Tools designed to analyze functional genomics data include XCluster for microarray analysis and Lutefisk and EMOWSE for proteomics analysis. XCluster (http://genetics. stanford. edu /~sherlock/cluster.html) generates clusters from gene expression data based on either SOM or hierarchical clustering. Lutefisk (Johnson and Taylor, 2002) is a proteomics tool designed to analyze mass spectra data, with considerable flexibility provided in the form of user-specified parameters for mass tolerance, spectral processing, etc. The results of this analysis can then be used to search selected protein databases for peptide sequence matches. The EMOWSE tool (http://www.hgmp.mrc.ac.uk/Software/EMBOSS/) is used to identify proteins based on their peptide mass fingerprint. By pasting in the masses of the tryptic peptides from the mass spectrum, and indicating the enzyme used for protein digestion, this analysis identifies proteins matching the spectrum in order of decreasing probability.

Data Mining and Queries

Numerous queries have been developed by the PlasmoDB team, based largely on inquiries from malaria researchers seeking to mine genomic-scale datasets for lists of genes that match specified parameters. For example, some users might want to search for all genes whose protein products are represented with a 3-D structure entry in PDB (Deshpande et al., 2005). Although this information is provided under the Protein view for each individual gene, looking through each individual Gene Page for this information is inefficient and impractical.

Data Mining Queries are accessible either as a series of pull-down menus on the PlasmoDB home page (Fig. 2), or via the Query link in the tool bar at the top of each page. On the Query page, options are organized under a variety of categories based on the data type and the kind of question to be asked. For example, in seeking putative transcription factors, users may query PlasmoDB to search the annotation text of *Plasmodium* genes, including GO terms and any uncurated comments provided by PlasmoDB users (if so desired). Users may also search for all genes that meet a particular set of (user-defined) criteria for chromosomal location, gene structure, gene features (e.g. signal peptides, transmembrane domains etc), the availability of crystal structures or structural models, presence in specified metabolic pathways, etc. Tools are also available for searching the results of microarray and proteomics data, and for identifying genes with a specific species distribution of putative orthologs.

In addition to conducting simple queries designed to identify (for example) all putative transcription factors, or all genes with >5 exons, or all genes that are transcribed in gametocytes, more complex questions can be formulated by combining queries, as illustrated in Figures 2 and 3. In this example, a user interested in identifying potential vaccine targets (Tongren et al., 2004) might wish to begin by searching for proteins located on the surface of the cell, as specified by the presence of a secretory signal sequence, a transmembrane domain, or both. A further criterion for inclusion might include protein presence in merozoite-stage parasites, on the assumption that these may be the most likely to be antigens in free merozoites. One might also be interested in genes that are conserved in other *Plasmodium* species, but absent from the human and mouse genome. Each of these queries is shown in Fig. 2: for example, this user has selected "Signal Peptide" from the "Sequence Feature" query menu, bringing up a page that may be configured based on the database and chromosome to be searched, and the algorithm of interest. Executing this query retrieves a long list of candidate secretory proteins (not shown, but see below).

All queries and the associated results generated in each PlasmoDB session are entered into the History page, accessible from the blue tool bar. As shown in Fig. 3, a simple signal sequence query identifies 654 hits, the expression query noted above yields 323 hits, and the phylogenetic query produces 1925 hits, when formulated as above. Taking the intersection of these queries (union and subtraction operations are also permitted) yields a total of 20 hits, including the leading vaccine antigens MSP1 and AMA1 and a variety of other genes that might be of interest to explore as candidate antigens.

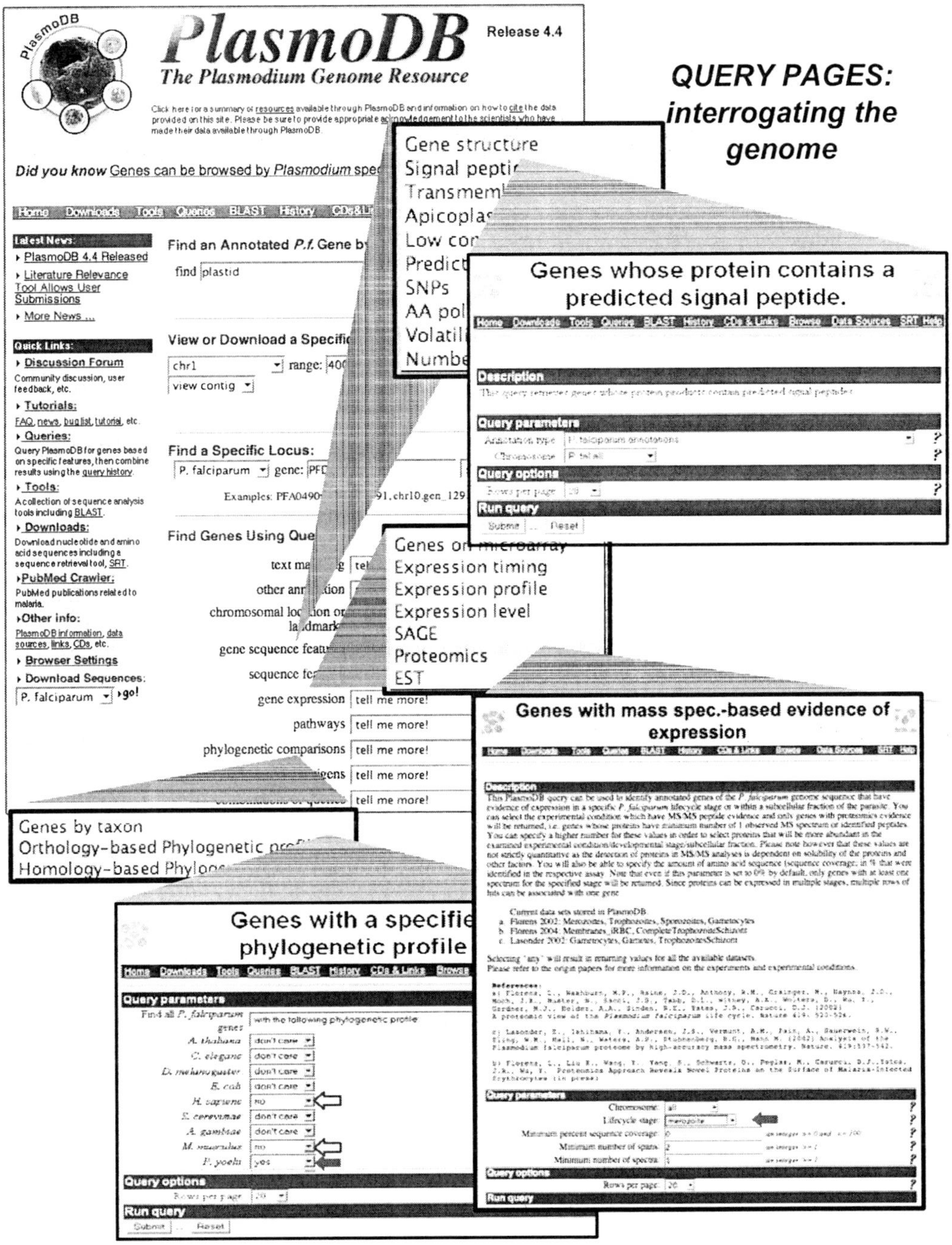

Figure 2. Gene Queries using PlasmoDB. A wide variety of dynamic queries may be formulated to interrogate the PlasmoDB database, using pull-down menus available on the home page (also accessible via the 'Queries' button on the blue tool bar, via the help and tutorial pages, and from links at relevant locations throughout the site). Three queries are shown, focused on characteristics that might be of interest to researchers seeking to mine the database for candidate vaccine antigens: (i) a search for genes predicted to encode a secretory signal sequence, (ii) a search for proteins that are present in merozoites, and (iii) a search for genes that are conserved in *P. yoelii,* but *not* in the human or mouse genomes.

Functional genomics data available at PlasmoDB includes genome-wide expression profiles for *P. falciparum* at various time points throughout the intraerythrocytic cycle, as well as (more limited) data for gametocytes and sporozoites (LeRoch et al., 2003; Bozdech et al., 2003). Data for different strains (HB3 vs. 3D7), synchronization protocols (sorbitol vs. temperature synchronization), and platforms (spotted oligonucleotide [glass slide] vs. photolithographic [Affymetrix] arrays), have been transformed so as to facilitate direct comparison between these results (cf. Fig. 1). While these datasets differ – chiefly in the high degree of time resolution available in the Bozdech dataset, vs. the large number of specific probes available in the LeRoch dataset – the high degree of concordance between data from these two groups lends high confidence to the results obtained. Protein expression data are also available in PlasmoDB, covering many developmental stages (Florens et al., 2002; Florens et al., 2004; Lasonder et al., 2002).

Several queries enable the user to identify genes that exhibit a desired expression profile, based on the profile of individual genes or user-specified profiles. For example, one researcher might wish to search for those genes whose expression in the glass slide dataset is closest to that of PFD0830w (DHFR-TS; see Fig. 1), or for genes whose expression matches a more complicated profile as drawn by the user on the Expression Profile query page. Alternatively, another might search for genes whose expression is maximal in late schizonts in the LeRoch dataset, or during the 40-48 hr interval in the Bozdech dataset. A third might seek genes for which experimental evidence supports expression of both RNA and protein in gametocytes.

A phylogenetic query allows PlasmoDB users to retrieve genes for which orthologs are conserved in (or absent from) species of specific interest. For example, an evolutionary biologist might wish to identify genes conserved in eukaryotes, but absent from prokaryotes. A drug developer might seek genes found in both *P. falciparum* and *P. yoelii,* but absent from the human (or murine) host (cf. Fig. 2). The volatility, or mutability, of a gene can also be queried, identifying cases where the protein sequence may have been more (or less) subject to positive selection (e.g. essential housekeeping genes) or negative selection (e.g. immunodominant genes) (Plotkin et al., 2004).

The results obtained from most types of queries conducted during a single session are automatically stored on the History page, accessible via the blue tool bar. Recent enhancements allow users to add individual genes to the History page by clicking "Add this gene to your History" on the Gene Page view (Fig. 1), and to delete query results no longer of interest. Applying Boolean set operators (union, intersection, subtraction) to these results enables the user to build more complex queries, as illustrated in Fig. 3. Results of interest (protein, mRNA, DNA sequence, chromosomal location, predicted transmembrane domains, GO annotations, etc) may then be extracted from the database using the ReportMaker tool. For example, the user might wish to download 1 kb of genomic sequence upstream of each gene in a defined set. Sequences can also be retrieved using the multiple sequence retrieval tool (SRT).

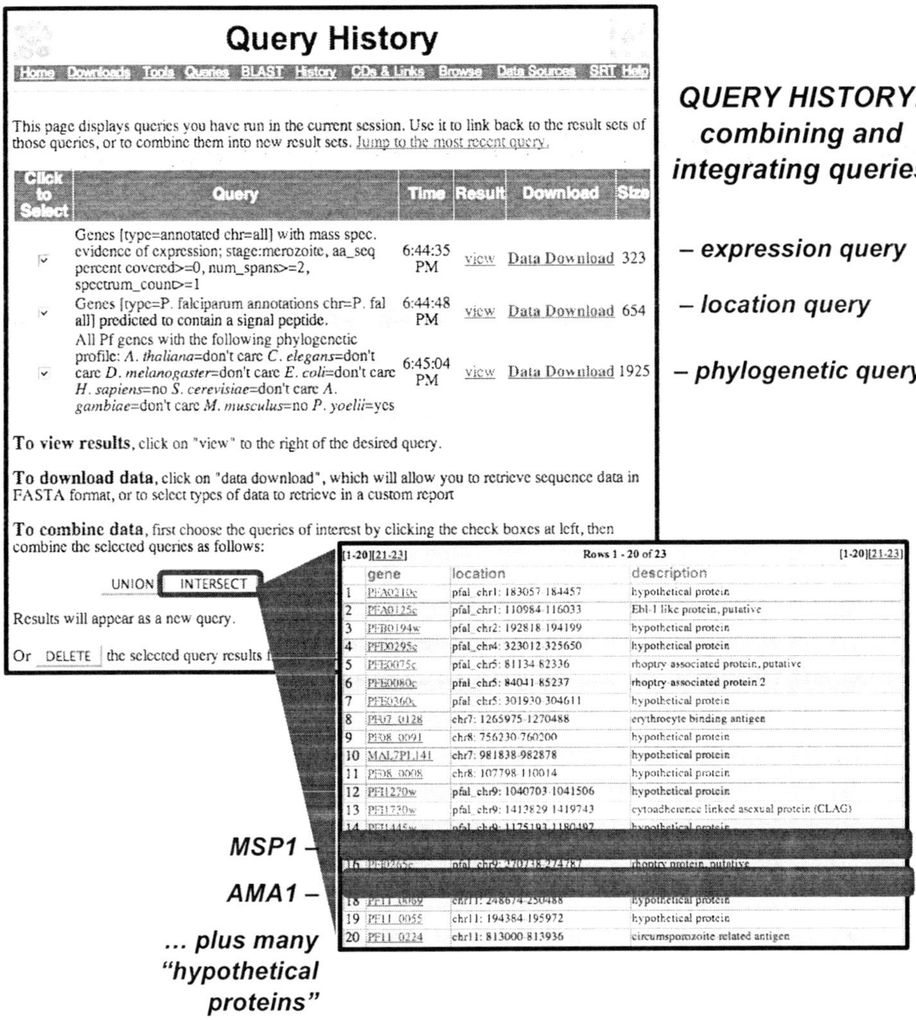

Figure 3. Using the 'History' function to integrate gene queries. The 'History' page, accessible via the blue tool bar at the top of each page, provides a list of all queries conducted during the current session. Using the boxes at left to specify individual queries of continuing interest permits these datasets to be combined (union), subtracted, or intersected. Intersecting the three queries defined in Fig. 2 yields 23 *P. falciparum* genes that satisfy the specified criteria with respect to phylogenetic distribution, expression, and predicted subcellular location. This list includes MSP1, AMA1, and numerous "hypothetical proteins" that may warrant further investigation as candidate vaccine antigens.

The Database behind PlasmoDB

The PlasmoDB database is managed using a relational database management system (RDBMS), implementing the Genomic Unified Schema (GUS) to store a variety of data types, including sequence and functional genomics data (Davidson et al., 2001). The design of GUS is based on the DNA→RNA→protein central dogma of molecular biology, using this principle to organize, store and integrate sequence and functional genomics data so that researchers can access the entirety of information available for individual DNA sequences, genes, and proteins (or groups of these entities). The GUS system is developed as an active open source project and is freely available at http://www.gusdb.org.

GUS is comprised of several schemas, each of which is modeled to store a specific type of data. The DoTS schema is designed to store sequence information, such as genomic sequence and EST data, while the RAD namespace is designed to store microarray data. The SRes (shared resource) namespace stores controlled vocabularies and ontologies, which are used for annotating the data and the Core namespace stores information about data provenance. In the latest release of GUS (v3.5), a Study schema for describing samples, design, and protocols for experimental studies and a Prot schema for describing liquid chromatography and mass spectrometry applications and results. As the DoTS schema is based on the central dogma, each step of the process from genomic DNA to protein is tracked, and the steps of this process annotated. Similarly, each step of a functional genomics experiment is tracked following guidelines put forward by standards groups (i.e., MIAME for micorarrays [Brazma et al., 2001], MIAPE for proteomics [Orchard et al., 2004]).

Data organization in the GUS schema is highly defined, or "strongly typed", in several ways. Concepts are modeled so that specific attributes are represented by individual columns in the relevant schema tables, and filled with the appropriate data. Tables and attributes are named such that they specify the type of data to be stored. For example, the DoTS schema table 'TranslatedAASequence' stores translated protein sequence data (as its name implies), and also contains specific columns for the protein sequence itself, and the sequence length. Keys are used to explicitly state the relationships between individual tables, such as those defining intron-exon structure, the resulting mRNA, and the predicted protein sequence. Such strong typing of the schema provides a measure of self-documentation, and allows researchers who use this system to share data with some degree of confidence that they know what they are getting.

Support and Feedback

The Support System – accessible at the bottom of every PlasmoDB web page – provides a forum for users to report bugs, post messages, ask questions, and view comments posted by others. Questions may be posted publicly (anonymously, if desired), enabling others who enter the forum to search for commonly asked questions and learn more about PlasmoDB. The system also enables users to communicate with the PlasmoDB development team privately, with questions, comments, and bug reports (describing the error encountered while using PlasmoDB, and the URL of the broken page). Each inquiry is entered into a tracking system, and assigned to the appropriate developer at PlasmoDB, so that answers may be provided directly to the individual user, or publicly in cases of general interest. Users are encouraged to take advantage of this feature to ask questions, report bugs, and make suggestions for improvements to the PlasmoDB database.

User-support features include a tutorial section, a mechanism for collecting and displaying user comments, and a PlasmoDB Support System. Annotations provided by curators at the genome sequencing centers (who are responsible for the official sequence records) are loaded into the database before each new release of PlasmoDB. The User

Comments feature now allows users to enter further comments at any time during a release cycle. In order to ensure proper attribution, registration is required before entering a User Comment. These comments are visible to all users of PlasmoDB via the Summary and Annotation gene page views, and may be queried using text search tools. User comments are also passed along to curators at the genome sequencing centers for review and possible inclusion in the official genome annotation.

A prototype implementation of natural language processing algorithms from the Mining the Bibliome project (R. MacDonald and F. Pereira, Univ. Pennsylvania, unpublished) provides links to published literature, ranked according to potential relevance. As this is a prototype tool, users are invited to rank the actual utility of these articles, so that this feedback may be employed to train future versions of the automatic article-annotation software.

Future Plans

As more *Plasmodium* species are sequenced, and additional functional genomics data is generated in the malaria field, this information will be incorporated into PlasmoDB. New genomic and EST data will be added for several *Plasmodium* species, including *P. vivax*, and this information will be available for queries similar to those now focused on *P. falciparum*. The database will also be configured so as to permit organism-specific views, so that the data may be approached from a *P. vivax*- or murine malaria-centric viewpoint. With support from the Bioinformatics Resource Center initiative of the US NIH (http://www.niaid.nih.gov/dmid/genomes/brc/default.htm), PlasmoDB will also be incorporated into an integrated Apicomplexan Parasite Database (ApiDB, http://apidb.org), including information on the related parasites *Toxoplasma* and *Cryptosporidium*.

New functional genomics datasets anticipated in the near future include: (*i*) numerous expression profiling studies, including analysis of additional life cycle stages, mutant parasite lines, drug-treated samples, and array studies on rodent malaria species; (*ii*) additional proteomics studies, focusing on various life cycle stages in several *Plasmodium* species; (*iii*) protein-protein interaction datasets, including the results from yeast two-hybrid studies, immunoprecipitation experiments, and computational predictions; (*iv*) additional protein structural data, and reagents available from structural genomics projects; (*v*) population diversity datasets, including whole-genome microsatellite studies; (*vi*) genome-scale analysis of subcellular localization (both experimentally-determined and computationally predicted); and (*vii*) genomic-scale analysis of predicted and experimentally determined indices of immunogenicity.

This integrated architecture will support queries based on cross-genome comparisons. For example, when asking for all genes expressed in a particular developmental stage, one will be able to integrate across multiple *Plasmodium* species. The ability to correlate the results from multiple experimental approaches will also be supported, providing (for example) a more detailed comparison of whole genome microarray data sets performed on different array platforms. PlasmoDB will also support comparisons of microarray and protein expression studies (cf. LeRoch et al., 2004).

PlasmoDB will continue to provide analysis methods aimed at predicting protein function, including the identification of orthologs and paralogs of proteins that have been annotated in other species (Li et al., 2003), phylogenetic profiling (Date and Marcotte, 2003), and ontology-based pattern identification (Zhou et al., 2004). In addition, these data will be incorporated into a framework for studying protein and metabolic pathway networks, enhancing our understanding of parasite biology and the identification of diagnostics, drugs, and vaccines.

Conclusion

PlasmoDB is designed as a resource for sequence and functional genomics data related to *Plasmodium* species. The underlying architecture exploits a relational database management system that implements the Genomics Unified Schema, enabling complex queries intended to elucidate the biology of malaria. Information on the genomes of various *Plasmodium* parasites may be gleaned by browsing the genome in Sequence View mode, and information on specific genes may be obtained from the various Gene Pages. Powerful analysis tools and user-defined queries allow researchers to formulate (and obtain answers to) their own questions.

The PlasmoDB database receives many thousands of 'hits' daily: >6 million in 2004, from >40 thousand unique users, in >100 countries worldwide. But this resource will only remain valuable to the extent that it keeps pace with the questions that researchers wish to ask. All of those using PlasmoDB are therefore encouraged to communicate their comments, questions, and concerns, using the Support System noted above. Scientists engaged in the production of genomic-scale datasets are encouraged to contact the PlasmoDB team as early as possible during the design of such experiments, to ensure that the results of their research can be accommodated, displayed, and queried as soon as they are ready for release.

Acknowledgements

The utility of PlasmoDB depends on the many researchers worldwide who generate the relevant data, and consent to make this data available to the community, often well in advance of formal publication. Particular thanks go to those engaged in the production of genomic-scale datasets, including researchers at the Sanger Institute, TIGR, and Stanford responsible for generating, assembling, annotating, and refining the primary sequence data. We are also grateful for financial support from the Burroughs Wellcome Fund and the National Institutes of Health (NIAID R01 AI058515 and NIAID Bioinformatics Resource Center contract HHSN266200400037C).

References

Altschul, S.F., Madden, T..J., Schaffer, A.A., Zhang, J., Zhang, Z., Miller, W. and Lipman, D.J. (1997) Gapped BLAST and PSI-BLAST: a new generation of protein database search programs. *Nucl. Acids Res.* 25, 3389-3402.

Bahl, A., Brunk, B.P., Coppel, R.L. , Crabtree, J., Diskin, S.J., Fraunholz, M.J., Grant, G., Gupta, D., Huestis, R.L., Kissinger, J.C., Labo, P., Li, L., McWeeney,S.K., Milgram, A.J., Roos, D.S., Schug, J. and Stoeckert, C.J., Jr. (2002). PlasmoDB: The *Plasmodium* genome resource. An integrated database providing tools for accessing and analyzing mapping, expression and sequence data (both finished and unfinished). *Nucl. Acids Res.* 30, 87-90.

Bahl, A., Brunk, B.P., Crabtree, J., Fraunholz, M.J., Gajria, B., Ginsburg, H., Grant, G.R., Gupta, D., Kissinger, J.C., Labo, P., Li, L., Mailman, M.D., Milgram, A.J., Pearson, D.S., Roos, D.S., Schug, J., Stoeckert, C.J., Jr. and Whetzel. P. (2003) PlasmoDB: The *Plasmodium* genome resource. Tools for integrating experimental and computational data. *Nucl. Acids Res.* 31, 212-215.

Bateman, A., Coin, L., Durbin, R., Finn, R.D., Hollich, V., Griffiths-Jones, S., Khanna, A., Marshall, M., Moxon, S., Sonnhammer, E.L., Studholme, D.J., Yeats, C., and Eddy, S.R. (2004) The Pfam protein families database. *Nucl. Acids Res.* 32, D138-41.

Bender, A., van Dooren, G.G., Ralph, S.A., McFadden, G.I. and Schneider, G. (2003) Properties and prediction of mitochondrial transit peptides from *Plasmodium falciparum*. *Mol. Biochem. Parasitol.* 132, 59-66.

Benson, D.A., Karsch-Mizrachi, I., Lipman, D.J., Ostell, J., and Wheeler, D.L. (2005) GenBank. *Nucl. Acids Res.* 33, D34-8.

Berry, A.E., Gardner, M.F., Caspers, G.J., Roos, D.S. and Berriman, M. (2004) Curation of the *Plasmodium falciparum* genome. *Trends Parasitol.* 20, 548-552

Bozdech, Z., Llinas, M., Pulliam, B.L., Wong, E.D., Zhu, J. and DeRisi, J.L. (2003) The transcriptome of the intraerythrocytic developmental cycle of *Plasmodium falciparum*. *PLoS Biol.* 1, 85-100.

Brazma, A., Hingamp, P., Quackenbush, J., Sherlock, G., Spellman, P., Stoeckert, C., Aach J., Ansorge W., Ball CA., Causton HC., Gaasterland T., Glenisson P., Holstege FC., Kim IF., Markowitz V., Matese JC., Parkinson H., Robinson A., Sarkans U., Schulze-Kremer S., Stewart J., Taylor R., Vilo J., and Vingron, M. (2001) Minimum information about a microarray experiment (MIAME)-toward standards for microarray data. *Nature Genetics* 29, 365-71.

Carlton, J. (2003) *Plasmodium vivax* genome sequencing project. *Trends Parasitol.* 19, 227-31.

Carlton, J., Silva, J. and Hall, N. (2005) The genome of model malaria parasites, and comparative genomics. *Curr. Issues Molec. Biol.* 7, 23-37.

Coppel, R.L., Roos, D.S. and Bozdech, Z. (2004) The genomics of malaria infection. *Trends Parasitol.* 20, 553-557.

Date, S.V., and Marcotte, E.M. (2003) Discovery of uncharacterized cellular systems by genome-wide analysis of functional linkages. *Nature Biotechnol.* 21, 1055-62.

Davidson, S., Crabtree, J., Brunk, B.P., Schug, J., Tannen, V., Overton, G.C. and Stoeckert, C.J. (2001) K2/Klesli and GUS: Experiments in integrated access to genomic data sources. *IBM Systems J.* 40, 512–531.

Deshpande, N., Addess, K.J., Bluhm, W.F., Merino-Ott, J.C., Townsend-Merino, W., Zhang, Q., Knezevich, C., Xie, L., Chen ,L., Feng, Z., Green, R.K., Flippen-Anderson, J.L., Westbrook, J., Berman, H.M. and Bourne, P.E. (2005)The RCSB Protein Data Bank: a redesigned query system and relational database based on the mmCIF schema. *Nucl. Acids Res.* 33, D233-7.

Doolan, D.L., Southwood, S., Freilich, D.A., Sidney, J., Graber, N.L., Shatney, L., Bebris, L., Florens, L., Dobano, C., Witney, A.A., Appella, E., Hoffman, S.L., Yates, J.R., 3rd, Carucci, D.J. and Sette, A. (2003) Identification of *Plasmodium falciparum* antigens by antigenic analysis of genomic and proteomic data. *Proc. Nat'l Acad. Sci. USA* 100, 9952-7.

Florens, L., Washburn, M.P., Raine, J.D., Anthony, R.M., Grainger, M., Haynes, J.D., Moch, J.K., Muster, N., Sacci, J.B., Tabb, D.L., Witney, A.A., Wolters, D., Wu, Y., Gardner, M.J., Holder, A.A., Sinden, R.E., Yates, J.R., and Carucci, D.J. (2002) A proteomic view of the *Plasmodium falciparum* life cycle. *Nature* 419, 520-526.

Florens, L., Liu, X., Wang, Y., Yang, S., Schwartz, O., Peglar, M., Carucci, D.J., Yates, J.R., and Wu, Y. (2004) Proteomics approach reveals novel proteins on the surface of malaria-infected erythrocytes. *Molec. Biochem. Parasitol.* 135, 1-11.

Foth, B.J., Ralph, S.A., Tonkin, C.J., Struck, N., Fraunholz, M.J., Roos, D.S., Cowman , A.F., and McFadden, G.I. (2003) Dissecting apicoplast targeting in the malaria parasite *Plasmodium falciparum. Science* 299, 705-708.

Gardner, M.J., Hall, N., Fung, E., White, O., Berriman, M., Hyman, R., Carlton, J., Pain, A., Nelson, J.E., Bowman, S., Paulsen, I.T., James, K., Eisen, J.A., Rutherford, K., Salzberg, S., Craig, A., Nene, V., Shallom, S., Suh, B., Peterson, J., Angiuoli, S., Pertea, M,,Allen, J., Selengut, J., Haft, D., Vaidya, A., Fairlamb, A., Roos, D.S., McFadden, G.I., Cummings, L.M., Mungall, C., Kanapin, A.A., Venter, J.C., Carucci, D.J., Hoffman, S.L., Newbold, C., Davis, R.W., Fraser, C.M., and Barrell, B. (2002) The genome sequence of the human malaria parasite *Plasmodium falciparum. Nature* 419, 498-511.

Hall, N., Karras, M., Raine, J.D., Carlton, J.M., Kooij, T.W., Berriman, M., Florens, L., Janssen, C.S., Pain, A., Christophides, G.K., James, K., Rutherford, K., Harris, B., Harris, D., Churcher, C., Quail, M.A., Ormond, D., Doggett, J., Trueman, H.E., Mendoza, J., Bidwell, S.L., Rajandream, M.A., Carucci, D.J., Yates, J.R., Kafatos, F.C., Janse, C.J., Barrell, B., Turner, C.M., Waters, A.P. and Sinden, R.E. (2005) A comprehensive survey of the *Plasmodium* life cycle by genomic, transcriptomic, and proteomic analyses. *Science* 307, 82-86.

Harris M.A., Clark, J., Ireland, A., Lomax, J., Ashburner, M., Foulger, R., Eilbeck, K., Lewis, S., Marshall, B., Mungall, C., Richter, J., Rubin, G.M., Blake, J.A,. Bult, C., Dolan, M., Drabkin, H., Eppig, J.T., Hill, D.P., Ni, L., Ringwald, M., Balakrishnan, R., Cherry, J.M., Christie, K.R., Costanzo, M.C., Dwight, S.S., Engel, S., Fisk, D.G., Hirschman, J.E., Hong, E.L., Nash, R.S., Sethuraman, A., Theesfeld, C.L., Botstein, D., Dolinski, K., Feierbach, B., Berardini, T., Mundodi, S., Rhee, S.Y., Apweiler, R., Barrell, D., Camon, E., Dimmer, E., Lee, V., Chisholm, R., Gaudet, P., Kibbe, W., Kishore, R., Schwarz,

E.M., Sternberg, P., Gwinn, M., Hannick, L., Wortman, J., Berriman, M., Wood ,V., de la Cruz, N., Tonellato, P., Jaiswal, P., Seigfried, T., and White, R. (2004) Gene Ontology Consortium. 2004. The Gene Ontology (GO) database and informatics resource. *Nucl. Acids Res.* 32, D258-61.

Hertz-Fowler, C., Peacock, C.S., Wood, V., Aslett, M., Kerhornou, A., Mooney, P., Tivey, A., Berriman, M., Hall, N., Rutherford, K., Parkhill, J., Ivens, A.C., Rajandream, M.A., and Barrell, B. (2004) GeneDB: a resource for prokaryotic and eukaryotic organisms. *Nucl. Acids Res.* 32, D339-43.

Hiller, N.L., Bhattacharjee, S., van Ooij, C., Liolios, K., Harrison, T., Estrano, C.L. and Haldar., K. (2004) A host-targeting signal in virulence proteins reveals a 'secretome' in malarial infection. *Science* 306, 1934-7.

Hulo N., Sigrist C..J.A., Le Saux V., Langendijk-Genevaux P.S., Bordoli L., Gattiker A., De Castro E., Bucher P., and Bairoch A. (2004) Recent improvements to the PROSITE database. *Nucl. Acids Res.* 32, D134-137.

Johnson, R.S. and Taylor, J.A. (2002) Searching sequence databases via *de novo* peptide sequencing by tandem mass spectrometry. *Molec. Biotechnol.* 22, 301-315.

Kaiser, K., Matuschewski, K., Camargo, N., Ross, J. and Kappe, S.H. (2004) Transcriptome profiling identifies *Plasmodium* genes encoding pre-erythrocytic stage-specific proteins. *Molec. Microbiol.* 51, 1221-32.

Kissinger, J.C., Brunk, B.P., Crabtree, J., Fraunholz, M.J., Gajria, B., Milgram, A.J., Pearson, D.S., Schug, J., Bahl, A., Diskin, S.J., Ginsburg, H., Grant, G.R., Gupta, D., Labo, P., Li, L., Mailman, M.D., McWeeney, S.K., Whetzel, P., Stoeckert , C.J., Jr. and Roos, D.S. (2002) The *Plasmodium* genome database: Designing and mining a eukaryotic genomics resource. *Nature* 419, 490-492.

Kissinger, J.C., and Roos, D.S. (2004) Getting the most out of bioinformatics resources. *In:* Malaria Parasites, A.P. Waters and C.J. Janse, editors. Horizon, Norfolk UK. *In press.*

Lasonder, E., Ishihama, Y., Andersen, J.S., Vermunt, A. M., Pain, A., Sauerwein, R.W., Eling,W.M., Hall, N., Waters, A.P., Stunnenberg, H.G. and Man, M. (2002) Analysis of the *Plasmodium falciparum* proteome by high-accuracy mass spectrometry. *Nature* 419, 537-542.

Le Roch, K.G., Zhou, Y., Blair, P.L., Grainger, M., Moch, J. K., Haynes, J.D., De la Vega, P., Holder, A.A., Batalov,S., Carucci, D.J. and Winzeler, E.A. (2003) Discovery of gene function by expression profiling of the malaria parasite life cycle. *Science* 301, 1503-1508.

Le Roch K.G., Johnson, J.R., Florens, L., Zhou, Y., Santrosyan, A., Grainger, M., Yan, S.F., Williamson, K.C., Holder, A.A., Carucci,D.J., Yates, J.R. and Winzeler, E.A. (2004) Global analysis of transcript and protein levels across the Plasmodium falciparum life cycle. *Genome Res.* 11, 2308-18.

Li, L., Stoeckert, C.J., Jr. and Roos, D.S. (2003) OrthoMCL: Identification of ortholog groups for eukaryotic genomes. *Genome Res.* 13, 2178-2190.

Li, L., Crabtree, J., Fischer, S., Pinney, D., Stoeckert , C.J., Jr., Sibley, L.D., and Roos, D.S. (2004) ApiESTDB: Analyzing clustered EST data of the apicomplexan parasites. *Nucl. Acids Res.* 32, 326-328.

Mamoun, C.B., Gluzman, I.Y., Hott, C., MacMillan, S.K., Amarkone, A.S., Anderson, D.I., Carlton, J.M., Dame, J.B., Chakrabarti, D., Martin, R.K., Brownstein, B.H. and Goldberg, D.E. (2001) Co-ordinated programme of gene expression during asexual intraerythrocytic development of the human malaria parasite *Plasmodium falciparum* revealed by microarray analysis. *Molec. Microbiol.* 39, 26-26.

Marti, M., Good, R.T., Rug, M., Knuepfer, E. and Cowman, A.F (2004) A unique export signal targets virulence and remodeling proteins from the malaria parasite to the host erythrocyte. *Science* 306, 1930-3.

McGuffin, L.J., Bryson, K. and Jones, D.T. (2000) The PsiPred protein structure prediction server. *Bioinformatics* 16, 404-405.

Milgram, A.J., Kissinger, J.C., Gajria, B., Pearson, D.S., Bahl, A., Labo, P. and Roos, D.S. (2003) *Plasmodium falciparum* GenePlot: Internet-independent access to the malaria parasite genome. *Nature* 422, CD-ROM (distributed in the March 6 issue).

Orchard S., Hermjakob, H., Julian, R.K. Jr., Runte, K., Sherman, D., Wojcik, J., Zhu, W. and Apweiler, R. (2004) Common interchange standards for proteomics data: Public availability of tools and schema. *Proteomics.* 4, 490-1.

Plotkin, J.B., Dushoff, J. and Fraser, H.B. (2004) Detecting selection using a single genome sequence of *M. Tuberculosis* and *P. falciparum. Nature.* 428, 942-945.

Ralph, S.A., van Dooren, G.G., Waller, R.F., Crawford, M.J., Fraunholz, M.J., Foth, B..J., Tonkin, C.J., Roos, D.S., and McFadden, G.I. (2004) Metabolic pathway maps and functions of the *Plasmodium falciparum* apicoplast. *Nature Rev. Microbiol.* 2, 203-216.

Roos, D.S., Crawford, M.J., Donald, R.G.K., Fraunholz, M., Harb, O.S., He, C.Y., Kissinger, J.C., Shaw, M.K. and Striepen, B. (2002) Mining the *Plasmodium* genome database to define organellar function: What does the apicoplast do? *Phil. Trans. Royal Soc. London B Biol. Sci.* 357, 35-46.

Sam-Yellowe T.Y., Florens, L., Wang, T., Raine, J.D., Carucci,D.J., Sinden, R. and Yates, J.R. (2004) Proteome analysis of rhoptry-enriched fractions isolated from *Plasmodium* merozoites. J *Proteome Res.* 3, 995-1001.

Schuler G. D. (1997) Sequence mapping by electronic PCR. *Genome Res.* 7, 541-50.

Servant, F., Bru, C., Carrere, S., Courcelle, E., Gouzy, J., Peyruc, D. and Kahn, D. (2002) ProDom: automated clustering of homologous domains. *Brief Bioinform.* 3, 246-251.

The Plasmodium Genome Database Collaborative. (2001) PlasmoDB: an integrative database of the *Plasmodium falciparum* genome. Tools for accessing and analyzing finished and unfinished sequence data. The Plasmodium genome resource. *Nucleic Acids Res.* 29, 66-69.

Tongren, J.E., Zavala, F., Roos, D.S. and Riley, E.M. (2004) Malaria vaccines: If at first you don't succeed … . *Trends Parasitol.* 20, 604-610

Zhou, Y., Young, J.A., Santrosyan, A., Chen, K., Yan, F.S. and Winzeler, E.A. (2005) *In silico* gene function prediction using ontology-based pattern identification. *Bioinformatics* 21, 1237-45.

In: In Silico Genomics and Proteomics
Editors: N. Mulder and R. Apweiler, pp. 201-217

ISBN 1-59454-995-8
© 2006 Nova Science Publishers, Inc.

Chapter 14

Integr8: Navigating Genome Reviews and the Proteome Analysis Database

Paul Kersey, Tamara Kulikova, Manuela Pruess and Rolf Apweiler
EMBL Outstation – The European Bioinformatics Institute
Wellcome Trust Genome Campus, Hinxton, Cambridge CB10 1SD, United Kingdom

Abstract

Genome Reviews and the Proteome Analysis Database both contribute to the Integr8 browser, which provides a platform for the interpretation of experimental and theoretical data in the biological context of the species. Genome Reviews is a database of annotated nucleotide sequences corresponding to the components of complete genomes. It aims to overcome some of the archival database problems by providing edited versions of entries representing complete genome sequences in the collaborative nucleotide sequence databases. The Proteome Analysis database provides statistical information about the proteomes of completely sequenced organisms, using data from different protein sequence resources that describe protein families and domains, protein clusters and protein functions. Together they provide the central data for Integr8, a web portal that offers integrated views of a variety of data concerning the biology of organisms with completely sequenced genomes. Among the features of Integr8 is the facility to analyse the composition of complete proteomes, both singly and in comparison with each other. The Integr8or, a tool for exploring the relationships between genes, provides a clear view of the relationships between genes, their neighbours, and the transcripts and proteins they encode. Additionally, using the resources of the CluSTr database, the Integr8or enables the user to identify potential orthologues and paralogues of each gene product, and to compare the genomic neighbourhood of related genes. The BioMart tool allows one to do customised queries over all proteomes, based on individual selections of species and protein attributes. These resources provide a powerful set of tools for the exploration of the biology of organisms whose genomes have been completely deciphered.

1. Introduction

Since the advent of whole genome sequencing in the mid-1990s, the sequences of over 220 cellular organisms have been completely determined, annotated, and deposited in the public repositories. The rate of deposition of such sequences is still increasing, with over 80 such genomes sequenced and made available during 2004. The availability of these data has enabled the development of new ways of interpreting information about individual genes and proteins in their biological context, and has underpinned the development of new experimental and theoretical fields such as transcriptomics, proteomics and systems biology. However, these new technologies have generated enormous quantities of data, meaning that the information needed to draw scientific conclusions is increasingly likely to be spread over many different primary resources. Coherently integrating such data, and offering access to it, has thus emerged as one of the most important challenges in bioinformatics.

Many technologies have been developed for providing views of dispersed data. Many employ a data warehousing approach, in which data is periodically imported from potentially many external resources into a common database schema, designed to support the optimal performance of common query types. Because query-optimised databases usually consist of a combination of common design motifs, tools for auto-generating customised user interfaces from such databases are also often available. In the biological sciences, the Sequence Retrieval System (Etzold *et al.*, 1996) is a widely used piece of software that utilises such an approach. An alternative strategy (used, for example, in the emerging technology known as Grid (Foster, 2003)) is to allow the user to directly query the raw resources, which may be distributed over many locations, but through an interface that hides the actual location of each resource. A layer of intelligent "middleware" receives the query, and calculates the most efficient strategy for retrieving and combining the individual data items requested by the user.

Computer software implementing these approaches can be of great use in the development of integrated databases; but there are a number of problems that these cannot address. Different resources do not necessarily maintain common identifiers for the data items they describe, or even agree on the definition of common terms and objects (Kersey *et al.*, 2003). If data items in different resources cannot be cross-correlated, then the ability to perform complex queries between different resources is lost. An additional problem is that different resources are updated on different update cycles, so that even if two resources are in agreement with each other at a certain point in time, it is not guaranteed that this will continue to be the case. As an illustration of this problem, one can consider there are currently 4,356 entries in the UniProt Knowledgebase (Bairoch *et al.*, 2005) that together define a non-redundant proteome set for the Gram-negative bacterium *Escherichia coli K-12*, representing the products of 4,396 genes. Of these, some 1,045 proteins (24%) have been assigned a protein sequence by UniProt curators other than that described in the EMBL/Genbank/DDBJ International Nucleotide Sequence Database (Benson *et al.*, 2005; Kanz *et al.*, 2005; Tateno *et al.*, 2005), to which the genome sequence was originally submitted in 1996 (Blattner *et al.*, 1997) but whose records were last updated in 1998.

A key point here is that EMBL, Genbank, and DDBJ function as archives for submitted sequences (which can, for example, then be referenced in publications): consistent with this, they allow submitters to retain ownership of their own data and in consequence, annotation of different entries is often not standardised in format or update frequency. With large scale genome sequencing projects, the problem caused by infrequent updates is especially severe, and much of the annotation provided on submitted genome sequences often consists of gene predictions; but as these are usually made by comparison with known sequences already in the databases, the original annotation may gradually appear less correct over time, even if there have been no advances in gene prediction algorithms or generation of new experimental data.

Genome Reviews (Kersey *et al.*, 2005), a database of annotated nucleotide sequences corresponding to the components of complete genomes, has been created to deal with this problem. In Genome Reviews, the annotation of EMBL entries is improved by importing more complete, up-to-date information from other data sources to supplement or replace the original annotation; the syntax and vocabulary of annotations are also standardised to improve inter-operability with other data sources. Genome Reviews also comprise one of the central data sources for Integr8 (Kersey *et al.*, 2005), a web portal that offers integrated views of a variety of data concerning the biology of organisms with completely sequenced genomes. Among the features of Integr8 is the facility to analyse the composition of complete proteomes, both singly and in comparison with each other.

2. The Genome Reviews

Sequence databases are comprehensive sources of information on nucleotide sequences and proteins and are universally used by molecular biologists and other life scientists. Normally such databases contain sequence data, annotation and reference information. The major resource for DNA and RNA sequences is the International Nucleotide Sequence Database (INSD), maintained collaboratively by the DNA Data Bank of Japan (DDBJ, Mishima, Japan) (Tateno *et al.*, 2005), the European Molecular Biology Laboratory (EMBL) Nucleotide Sequence Database from the European Bioinformatics Institute (EBI, Hinxton, UK) (Canz *et al.*, 2005), and GenBank at the National Center for Biotechnology Information (NCBI, Bethesda, USA) (Benson *et al.*, 2005). These nucleotide sequence databases are typical primary databases and are acting as data repositories, accepting, storing and making the data derived by sequencing freely available to the scientific community. Each nucleotide sequence submitted to the databases will be annotated to describe the experimental results and the interpretation of the results that the submitting sequencing group attach to them. The extent of the annotation in each entry depends very much on the aim of the particular research during which the data were obtained, the methods employed, the data type, etc. Nucleotide sequence data collected and stored in the databases are produced by genome sequencing projects (which account for the bulk of the data in the databases) and as a result of smaller sequencing efforts. DDBJ/EMBL/GenBank strives for completeness, with the aim of recording and making available every publicly known nucleic acid sequence. Due to the archival nature of the database, the final editorial control over a database entry in DDBJ/EMBL/GenBank rests with the submitters of individual records. That leads to a situation where many sequences stay unupdated forever and may therefore forever contain incorrect or incomplete information. Errors can be propagated to newly created entries, as an annotation of predicted genes is often inferred by similarity to the existing sequences; newly created erroneous annotation then appears to support the older instance of the same annotation,. This problem can only be overcome by active curation, which is impossible in archival datasets.

It must be noted that there is an important difference between primary (archival) and secondary (curated) databases. Primary databases represent results of experiments in the lab, supplemented with the interpretation of the results supplied by the research group. Secondary databases add value to the archival data by offering curated reviews of the primary data. Secondary databases are therefore a better data source for more stable and reliable data; primary databases can be used to monitor ongoing research and "history" of the research.

The fact that most of the data in DDBJ/EMBL/GenBank records cannot be updated, corrected or extended without the permission of the original submitter, and the redundancy of sequence data, has led to the creation of secondary nucleotide sequence databases. One of those is EMBL Genome Reviews, a database in EMBL-like format in which annotation has been corrected and extended (compared to the original submission) through the integration of data from many sources. The Genome Reviews database aims to overcome some of the archival database problems by providing edited versions of entries representing complete genome sequences in the collaborative nucleotide sequence databases. Each Genome Review represents an edited version of the original flatfile (in extended, EMBL compatible format, see below), maintaining the link with the primary submission via the use of identifiers, and with additional annotation imported from other data sources such as the UniProt Knowledgebase, InterPro (Mulder *et al.*, 2005) and others. For each particular data type a preferred source of information (database) is chosen and the annotation from that source is imported into the Genome Review to either supplement or to correct the annotation that existed in the original submission. The UniProt Knowledgebase, which constitutes a well-annotated resource where redundant information is merged, and where literature-based curation is used to extract experimental data, supplemented by manually confirmed results from various sequence analysis programs, has proved to be especially useful. In addition, annotations applied inconsistently among the original submissions have been standardised, and deleted in some cases, usually where the annotation coverage is low or varies significantly between genomes. In come cases CDS (CoDing Sequence) features, which have been identified by the UniProt curators as "false", i.e. unlikely to encode a real protein, have been removed. Plans for further annotation enhancement in Genome Reviews include introducing features (such as CDS or tRNA features) that are missing in the original submission.

Although new types of features and feature qualifiers have been introduced to describe additional data not previously present in EMBL (for example "/biological_process" and "/cellular_component" qualifiers), the overall number of feature and qualifier types decreased in comparison to the original submissions due to standardisation of the annotation (Kersey *et al.*, 2005) and to removal of redundant data. In addition to that, new features have been added to Genome Reviews by mapping features annotated on protein sequences onto corresponding regions of DNA. For example, regions of DNA encoding the mature peptides produced after cleavage of the primary translations, "mat_peptide" features, have been introduced. The number of cross-references (described by db_xref qualifiers) has been increased manifold in comparison with the original submissions.

The most notable change to feature qualifiers in comparison to original EMBL format has been the introduction of evidence tags. Evidence tags are attached to most feature qualifiers indicating the primary source of the information. The evidence tags show the database that was used as a source of the information and the identifiers within the database, where appropriate. The presence of an exclamation mark (!) before the database identifier indicates that a deduction has been made from the absence of this identifier from the database in question.

Evidence tags are always located at the end of the qualifier value. They are contained within curly braces (i.e. between the '{' character and the '}' character) and preceded by a space. Figure 1 shows an excerpt of a Genome Reviews sample record, specifically an example of CDS annotation in one of Genome Reviews, where the application of evidence tags can be seen clearly.

```
FT CDS      complement(139..591)
FT          /gene="SF3822 {UniProt/TrEMBL:Q6LW57}"
FT          /locus_tag="PBPRA0001 {UniProt/TrEMBL:Q6LW57}"
FT          /product="Hypothetical MioC homolog
FT          {UniProt/TrEMBL:Q6LW57}"
FT          /function="FMN binding {GO:0010181}"
FT          /function="electron transporter activity {GO:0005489}"
FT          /function="oxidoreductase activity {GO:0016491}"
FT          /biological_process="electron transport {GO:0006118}"
FT          /protein_id="CAG18456.1 {EMBL:CR354531}"
FT          /db_xref="GO:0005489 {GOA:Q6LW57}"
FT          /db_xref="GO:0006118 {GOA:Q6LW57}"
FT          /db_xref="GO:0010181 {GOA:Q6LW57}"
FT          /db_xref="GO:0016491 {GOA:Q6LW57}"
FT          /db_xref="HSSP:1AMO {UniProt/TrEMBL:Q6LW57}"
FT          /db_xref="InterPro:IPR001094 {UniProt/TrEMBL:Q6LW57}"
FT          /db_xref="InterPro:IPR008254 {UniProt/TrEMBL:Q6LW57}"
FT          /db_xref="UniParc:UPI000035B799 {EMBL:CAG18456}"
FT          /db_xref="UniProt/TrEMBL:Q6LW57 {EMBL:CR354531}"
FT          /codon_start=1
FT          /transl_table=11
FT          /translation="MSNITLITGSTLGGAEYVADHLSELLEQDGHSTEVINHANLSELN
FT          IDSIWLFVCSTHGAGDFPDNIQPFISQLTAQKPDLSALKYGVIGLGDSSYDTFCAAGII
FT          IDNLLNSLHAMKLGERLDIDISQHSVPEDAAESWFVEWKKHLCSKN"
```

Figure 1.Excerpt of a Genome Reviews sample record, showing an example of CDS annotation, where the application of evidence tags can be seen clearly.

The latest release of the EBI Genome Reviews at the time of writing (January 2005) contains a total of 335 files for each chromosome and plasmid from 196 prokaryotes with completely sequenced genomes. In total the Genome Reviews release files contain 629 million nucleotides and 585 thousand annotated protein-coding sequences. More up-to-date statistics are available from http://www.ebi.ac.uk/GenomeReviews/stats/. In later releases of Genome Reviews, data will also be available for eukaryotic organisms.

The Genome Reviews data is synchronised with the UniProt Knowledgebase and is distributed in EMBL-compatible format to ensure compatibility with existing tools. Genome Reviews are made available at http://www.ebi.ac.uk/GenomeReviews, via anonymous ftp at ftp://ftp.ebi.ac.uk/pub/databases/genome_reviews/. Supporting software is available via anonymous ftp.

3. The Integr8 Browser

The Integr8 web portal (Kersey *et al.*, 2005) (http://www.ebi.ac.uk/integr8) has been developed to offer a consistent view of genes, their transcripts and their proteins in their genomic context. The focus is on species whose genomes have been completely sequenced, and a focal point through which relevant data can be identified and downloaded. For each organism, Integr8 provides an overview of its biology and a detailed statistical analysis of its genome and proteome, represented using both textual and graphical information. Comparative analysis between genomes is also supported (Figure 2 shows an example for Genome Statistics for the fission yeast *Schizosaccharomyces pombe*). Additionally, users can customise their own analysis using BioMart, a powerful query tool originally developed to support access to the Ensembl database (Kasprzyk *et al.*, 2004) but whose coverage has since been extended to include data not only from Integr8, but also the European Macromolecular Structure Database (Velankar *et al.*, 2005), VEGA (Ashurst *et al.*, 2005) and dbSNP (Sherry *et al.*, 2001).

Integr8 is supported by 3 main data sources:

i)   Genome Reviews
ii) Non-redundant sets of UniProt entries representing each complete proteome. For prokaryotic organisms, these are constructed according to the HAMAP (High-quality Automated and Manual Annotation of microbial Proteomes) specifications (Gattiker *et al.*, 2003) used in the annotation of the UniProt Knowledgebase. Sets are also available for eukaryotic organisms, prepared by filtering the UniProt Knowledgebase using information from the EMBL Nucleotide Sequence Database and certain model organism databases. In some species (like human) where the level of multiple submissions is very high, additional entries are filtered out according to sequence similarity.
iii) IPI (the International Protein Index) (Kersey *et al.*, 2004). IPI provides comprehensive protein sets for certain higher metazoan species, by combining data from the UniProt Knowledgebase, Ensembl (Hubbard *et al.*, 2005) and RefSeq (Pruitt *et al.*, 2005) in a non-redundant fashion.

For each proteome set, additional information has been integrated from other resources such as HAMAP, InterPro, CluSTr (Kriventseva *et al.*, 2003), the Gene Ontology Annotation database (GOA) (Camon *et al.*, 2004) and others.

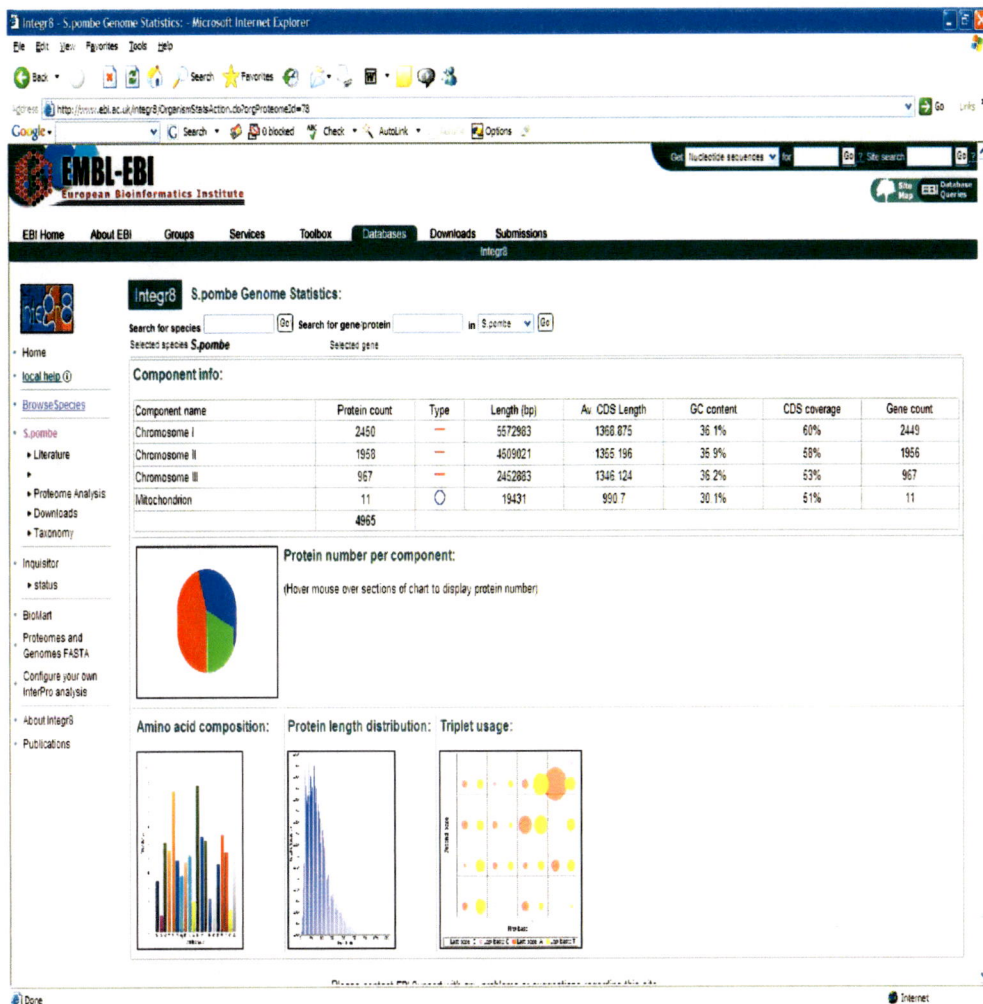

Figure 2. Genome Statistics for the fission yeast *Schizosaccharomyces pombe*, as represented in the Integr8 browser

Gene and Species Search

A simple search form provides access to summary information relevant to each species. This information includes a description, a list of recent publications, a list of the components of its genome, and information about the composition of these components (such as their length, average GC content, and the length and codon usage in the CDSs they contain), represented textually or graphically as appropriate. The search facility can also be used to search for proteins belonging to the non-redundant proteome sets, and the genes that encode them.

The Integr8or

The Integr8or is a new tool developed within the Integr8 portal for exploring the relationships between genes. Genes are related to each other according to their locations on the genome; but are also linked through the sequence, structure and function of the proteins they encode. Identification of the presence (or absence) of equivalent proteins in different species can shed light on the evolutionary history and functional characteristics of these species, and of the proteins themselves.

The Integr8or also offers a clear view of the different transcripts encoded by different genes; of which sequences (in the major DNA and protein sequence databases) correspond to each transcript; of which transcripts have been identified through the sequencing of mRNA molecules and which are only predictions from genomic DNA; and of the individual functional classification of each protein product. This view is generated by merging information derived from the UniProt Protein Knowledgebase (and through this resource, model organism databases such as FlyBase (Drysdale *et al.*, 2005) and WormBase (Chen *et al.*, 2005)), the UniParc Protein Sequence Archive (Leinonen *et al.*, 2004), the EMBL Nucleotide Sequence Database, and, through IPI, additional transcripts and proteins derived from Ensembl and RefSeq (an overview of the results of this merging process is given in Table 1). Through the incorporation of data from many sources, Integr8 is able to offer a gene-centric view of the biology of an organism. As part of this view, each gene is shown in the context of its neighbours on the chromosome or plasmid that encodes it.

Table 1.Summary of data available for access via the
Integr8 web portal (Figures correct as of 1st January 2005)

Number of species represented in Interg8	224
Number of genes represented in Integr8	843,124
Number of transcripts represented in Integr8	892,377
Number of transcripts represented in Integr8 supported by the existence of mRNA sequence	79,280
Number of potential protein orthologues* identified	32,409,403

* most similar sequences to a protein in one species from another species

The real power of the Integr8or lies in its ability to allow users to navigate between views of closely related genes. Using information derived from the CluSTr database, which stores the pairwise similarity of all protein sequences present in the UniProt proteome sets and IPI, the most similar other sequences to each sequence have been identified. These sequences have then been categorised as potential orthologues (related sequences, diverged through evolutionary time following a speciation event) and potential paralogues (related sequences in the same species, separated by a gene duplication event). For each protein sequence, Integr8 offers access to the top-ranked potential orthologues and paralogues. Additionally, Integr8 allows the user to display a comparative view of the genes (and their genomic neighbourhoods) that encode these similar protein sequences. Gene order is frequently conserved in closely related species, and therefore a comparison of the neighbourhood of related genes can provide important evidence for establishing the existence of genuine homology (i.e. the existence of a common ancestral sequence). For known homologues, on the other hand, the study of the differences in their genomic neighbourhoods may provide insights into the evolutionary history of their species. An example, showing a comparison between the neighbourhoods of the L-fucolokinase gene in *Escherichia coli* and a number of related species, is shown in Figure 3.

To support the comparison of different regions of DNA, centred on potentially homologous genes, the Integr8or colours its display of each gene according to the domain architecture of the protein it encodes, according to the InterPro database. This makes it possible to see strong patterns of synteny even between bacterial species whose genes and proteins may be poorly annotated and whose annotated names therefore provide no clue to the function. Integr8's schematic view of genome organisation is especially well suited to the study of bacterial genomes, which have a high gene density but generally simple gene structures.

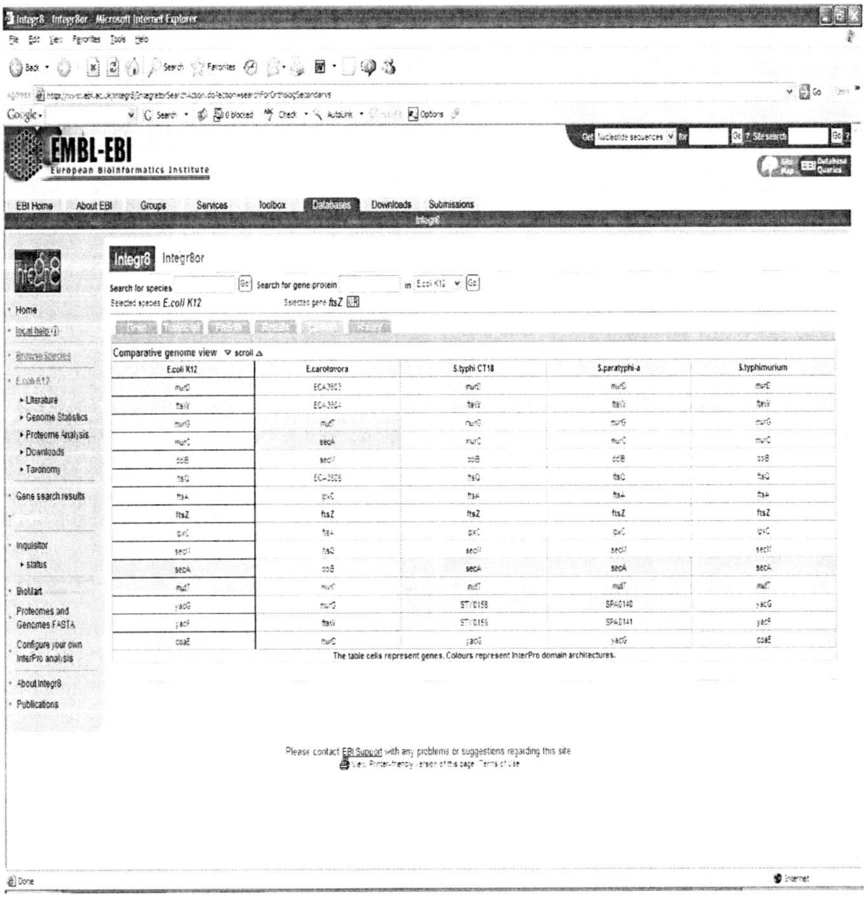

Figure 3. Using Integr8 to identify patterns of genomic synteny around potential orthologues. The display in the right hand panel shows the neighbourhood of the selected gene (L-fuculokinase in *E. coli*) and that of its potential orthologues in a number of (user-selected) species. The colours represent InterPro domain architecture, allowing the identification of probable conservation even where gene nomenclature is not conserved.

Sequence Similarity Search

The UniProt proteome sets are also available (at http://www.ebi.ac.uk/fasta33/proteomes) for sequence similarity searches, allowing users to restrict their analysis of a single sequence to the proteome set of their choice. IPI data sets can also be searched by FASTA (at http://www.ebi.ac.uk/fasta33/) or BLAST (http://www.ebi.ac.uk/blast).

Downloadable Resources

The following data is available for download from the Integr8 FTP site: (i) Genome Reviews files, (ii) UniProt complete proteome sets, (iii) IPI data sets, (iv) Files of InterPro matches for all proteins in each proteome set, (v) Files of Gene Ontology annotations for all proteins in each proteome set, (vi) "Chromosome tables", summary files mapping proteins represented in UniProt to their genomic locations. Users can additionally customise their own data for download using BioMart.

4. Proteome Analysis

The statistical analysis of proteomes forms a core part of the Integr8 web portal. It complements the Genome Reviews, in which whole genomes are annotated, by providing comparative information about the proteins encoded by the genomes, the so-called 'proteomes'. The proteome sets are built from the UniProt Knowledgebase, a protein sequence database that provides reliable, well-annotated data as the basis for the analysis. Proteome analysis data is available for all the completely sequenced organisms present in the Knowledgebase, spanning archaea, bacteria and eukaryota. These proteomes represent, naturally, only a static view on the set of encoded proteins, not taking into account the variations that can be found in different tissues, cell types, at different stages of development, under different environmental conditions etc. However, they are already useful for characterising an organism on a molecular biological basis. Many of these proteins usually are not described in detail so far, but some are, and for a number of the rest it is at least possible to determine with the help of sequence comparisons to which protein group or family they may belong, and thus what their functions may be. So the total of the proteins can already be analysed *in silico*, leading to an organism-specific 'profile' or composition of the proteome. Such a profile can give information about the potential functions of an organism, and enable comparisons between different organisms.

To analyse each organism's proteome, and to make this information publicly available, the Proteome Analysis database (Pruess *et al.*, 2003) has been developed. A highly integrative tool in itself, it has now been completely integrated into the Integr8 browser, described above, where it provides statistical information about the proteomes. To compile this body of information, data from different sources is used, dealing with the identification of protein families and domains, the clustering of proteins and the annotation of protein function and localisation.

One of these sources is the InterPro database, already mentioned before, an integrated documentation resource of protein families, domains and functional sites. InterPro includes the member databases PROSITE (Hulo *et al.*, 2004), Prints (Attwood *et al.*, 2003), Pfam (Bateman *et al.*, 2004), ProDom (Servant *et al.*, 2002), SMART (Letunic *et al.*, 2004), TIGRFAMs (Haft *et al.*, 2003), PIRSF (Wu *et al.*, 2003), and SUPERFAMILY (Gough *et al.*, 2001), all of which use protein signatures for identifying related proteins or elucidating protein function. They apply different methods for creating these signatures, most of which initiate from multiple sequence alignments of proteins known to belong to the family of interest. InterPro is an attempt to rationalise the different protein signatures into a comprehensive resource where those signatures designed to find the same protein family or domain are integrated into single InterPro entries. Accordingly, each entry contains one or more signatures from the individual member databases which all describe the same group of proteins. This enables researchers to use all the major protein signature databases at once, receiving the results in a single format, thus drawing on the strengths of all the databases and at the same time compensating for any downfalls they may have. This comprehensive knowledge of protein families and domains makes InterPro a powerful tool for describing the physical composition of a complete proteome and for inferring the biological consequences of this. Through Integr8, InterPro can be used to discover the most common domains, families and functional sites present in each proteome, find the proteins with the largest number of different InterPro classifications, and determine the overall coverage of the proteome by InterPro methods (unclassified proteins are usually completely uncharacterised, and therefore are either novel or wrongly predicted). The complete set of all matches between InterPro methods and sequences in each proteome set is also available for download. In addition, precomputed comparisons are available for interesting combinations of species, allowing users to see the differential representation of InterPro matches between them. Additional comparisons can be customised by the user and generated interactively.

Another source used for analysing the proteomes is the CluSTr database. It provides the possibility to cluster proteins into groups of related ones, by means of automatic classification. The clustering of UniProt Knowledgebase proteins is based on analysis of all pairwise comparisons between protein sequences using the Smith–Waterman algorithm (Smith and Waterman, 1981). Statistical significance is estimated using Monte-Carlo simulation resulting in a Z-score (Comet *et al.*, 1999). Analysis carried out at different levels of protein similarity yields a hierarchical organisation of clusters. By working with clusters at different levels of similarity, biologically meaningful clusters can be selected for different groups of proteins, making use of the flexibility of the database. Clusters for mammalian proteins, plant proteins and for 11 complete eukaryotic genomes (including *Homo sapiens*) have been built. In the proteome analysis part of the Integr8 browser, CluSTr covers currently these 11 complete eukaryotic genomes, with more to come. It also provides links to InterPro with its wealth of information on protein families, domains and functional sites.

A functional classification of proteomes is performed according to the assignment of proteins to Gene Ontology (GO) terms. This dynamic controlled vocabulary can be applied to all organisms, even while knowledge of gene and protein roles in cells is still accumulating and changing (Harris *et al.*, 2004). For the proteome analysis, not the whole range of GO terms is used: GO is very comprehensive with currently more than 17,350 terms in total, and this number is growing constantly (visit http://www.geneontology.org/ for more information). Instead, for assigning proteins to all GO terms, only a specially choosen selection of high level terms from each of the three Gene Ontology sections (molecular function, biological process and cellular component), called GO Slim, is used. (There are a number of such 'GO Slims' existing, each built to purpose for certain projects.).

A functional classification of the proteins within each proteome set has been generated to show the percentage of proteins involved in certain functions. All organisms are linked to the NEWT taxonomy browser (http://www.ebi.ac.uk/newt/index.html) and to related resources.

To complete the information about the proteomes, links to structural information databases like the Homology derived Secondary Structure of Proteins (HSSP) database (Dodge *et al.*, 1998), the Protein Data Bank (PDB; Berman *et al.*, 2003), and the Structural Classification of Proteins (SCOP) database (Lo Conte *et al.*, 2002) are provided, for individual proteins from each of the proteomes.

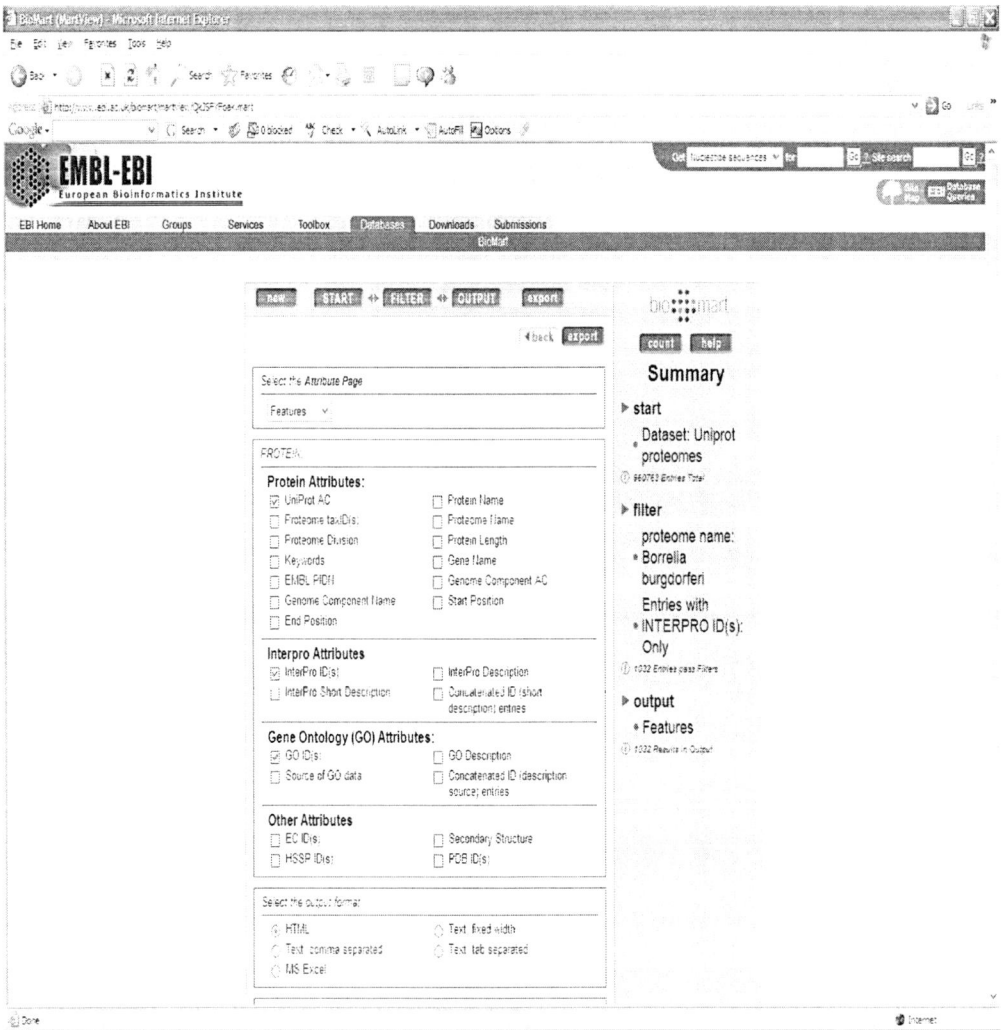

Figure 4. The BioMart interface: an example of how the selection of protein attributes can be done.

Another tool for viewing the contents of Integr8 and the proteome analyses is BioMart, a development of EnsMart (Kasprzyk *et al.*, 2004). Here the user can customise their own analysis. In a first step, after having chosen the 'Focus: UniProt Proteomes' (BioMart also offers data from the European Macromolecular Structure Database, VEGA and dbSNP to choose from), the user can do an individual selection. It is possible to choose one or more species, and then more filters can be applied like chromosomal location, the availability of types of external identifiers (like InterPro or GO IDs, gene names or keywords), and the assignment of particular GO terms. In a next step, different protein attributes can be chosen. As a result, the user will get a list of proteins that fulfill the demanded criteria. (Figure 4 shows an example for BioMart usage and different search attributes that can be chosen.)

Currently (January 2005), proteome analysis in Integr8 is available for 218 complete genomes.

5. Conclusion

The challenges of data integration are an increasingly important problem in bioinformatics, as increasing quantities of high throughput data are generated by new technologies. Such technologies have made possible the development of moves towards the understanding of complete organisms at a systems level; but the successful interpretation of primary results is highly dependent on the compatibility of data derived from different experiments, and of the inter-operability of different systems that make that data available.

Genome Reviews and Integr8 are two related projects that make a contribution to providing a platform for the interpretation of experimental and theoretical data in the biological context of the species. In Genome Reviews, the records submitted by genome sequencers to the nucleotide sequence repositories are available in updated form, with the scope of annotations extended and a regular annotation pipeline established to import up-to-date information; this is especially important as the original annotation is often largely inferred by comparison with other known data, and so can become dated even if the method of inference remains valid. With all imported data clearly sourced through the use of evidence tags, the standardisation of syntax and the addition of missing data (for example, for unannotated protein and tRNA-coding genes), Genome Reviews provides a solid platform for investigating the biology of complete genomes.

The Integr8 application allows users to interpret the data in Genome Reviews, and in other data sources, such as the UniProt Knowledgebase and IPI, through an interface that allows users to explore relationships between different types of biological objects, centred on individual genes. From a starting point of a single gene, Integr8 allows users to view that gene's neighbours, the transcripts it dictates and the proteins it encodes, and then to jump to genes of putative homology in selected species. The comparative view allows the user to compare schematic representations of the neighbourhoods of putative homologues, and thereby to identify genomic synteny through the architecture of the proteins described by potentially corresponding genes. Integr8 also contains substantial quantities of analytical data, summarising the genome and proteome of each species, and providing tools for the comparison of the proteomes from different species.

The data from Genome Reviews and Integr8 is also available in BioMart, a powerful interactive query tool that allows users to perform their own statistical analysis of the underlying data, and to download customised data combining input from the different primary sources used to build them.

Together, these resources provide a powerful set of tools for the exploration of the biology of organisms whose genomes have been completely deciphered; and a potential anchor point for the attachment of new types of data (that may become available to public repositories in the future) to existing resources such as the UniProt Knowledgebase.

Acknowledgement

This work was funded by the European Commission TEMBLOR grant, contract-no. QLRI-CT-2001-00015, under the RTD programme "Quality of Life and Management of Living Resources".

References

Ashurst, J. L., Chen, C. K., Gilbert, J. G., Jekosch, K., Keenan, S., Meidl, P., Searle, S. M., Stalker, J., Storey, R., Trevanion, S., Wilming, L. and Hubbard, T. (2005). The Vertebrate Genome Annotation (Vega) database. *Nucleic Acids Res,* 33 Database Issue, D459-465.

Attwood, T. K., Bradley, P., Flower, D. R., Gaulton, A., Maudling, N., Mitchell, A. L., Moulton, G., Nordle, A., Paine, K., Taylor, P., Uddin, A. and Zygouri, C. (2003). PRINTS and its automatic supplement, prePRINTS. *Nucleic Acids Res,* 31, 400-402.

Bairoch, A., Apweiler, R., Wu, C. H., Barker, W. C., Boeckmann, B., Ferro, S., Gasteiger, E., Huang, H., Lopez, R., Magrane, M., Martin, M. J., Natale, D. A., O'Donovan, C., Redaschi, N. and Yeh, L. S. (2005). The Universal Protein Resource (UniProt). *Nucleic Acids Res,* 33 Database Issue, D154-159.

Bateman, A., Coin, L., Durbin, R., Finn, R. D., Hollich, V., Griffiths-Jones, S., Khanna, A., Marshall, M., Moxon, S., Sonnhammer, E. L., Studholme, D. J., Yeats, C. and Eddy, S. R. (2004). The Pfam protein families database. *Nucleic Acids Res,* 32 Database issue, D138-141.

Benson, D. A., Karsch-Mizrachi, I., Lipman, D. J., Ostell, J. and Wheeler, D. L. (2005). GenBank. *Nucleic Acids Res,* 33 Database Issue, D34-38.

Berman, H., Henrick, K. and Nakamura, H. (2003). Announcing the worldwide Protein Data Bank. *Nat Struct Biol,* 10, 980.

Blattner, F. R., Plunkett, G., 3rd, Bloch, C. A., Perna, N. T., Burland, V., Riley, M., Collado-Vides, J., Glasner, J. D., Rode, C. K., Mayhew, G. F., Gregor, J., Davis, N. W., Kirkpatrick, H. A., Goeden, M. A., Rose, D. J., Mau, B. and Shao, Y. (1997). The complete genome sequence of Escherichia coli K-12. *Science,* 277, 1453-1474.

Camon, E., Barrell, D., Lee, V., Dimmer, E. and Apweiler, R. (2004). The Gene Ontology Annotation (GOA) Database--an integrated resource of GO annotations to the UniProt Knowledgebase. *In Silico Biol,* 4, 5-6.

Chen, N., Harris, T. W., Antoshechkin, I., Bastiani, C., Bieri, T., Blasiar, D., Bradnam, K., Canaran, P., Chan, J., Chen, C. K., Chen, W. J., Cunningham, F., Davis, P., Kenny, E., Kishore, R., Lawson, D., Lee, R., Muller, H. M., Nakamura, C., Pai, S., Ozersky, P., Petcherski, A., Rogers, A., Sabo, A., Schwarz, E. M., Van Auken, K., Wang, Q., Durbin, R., Spieth, J., Sternberg, P. W. and Stein, L. D. (2005). WormBase: a comprehensive data resource for Caenorhabditis biology and genomics. *Nucleic Acids Res,* 33 Database Issue, D383-389.

Comet, J. P., Aude, J. C., Glemet, E., Risler, J. L., Henaut, A., Slonimski, P. P. and Codani, J. J. (1999). Significance of Z-value statistics of Smith-Waterman scores for protein alignments. *Comput Chem,* 23, 317-331.

Dodge, C., Schneider, R. and Sander, C. (1998). The HSSP database of protein structure-sequence alignments and family profiles. *Nucleic Acids Res,* 26, 313-315.

Drysdale, R. A., Crosby, M. A., Gelbart, W., Campbell, K., Emmert, D., Matthews, B., Russo, S., Schroeder, A., Smutniak, F., Zhang, P., Zhou, P., Zytkovicz, M., Ashburner, M., de Grey, A., Foulger, R., Millburn, G., Sutherland, D., Yamada, C., Kaufman, T., Matthews, K., DeAngelo, A., Cook, R. K., Gilbert, D., Goodman, J., Grumbling, G., Sheth, H., Strelets, V., Rubin, G., Gibson, M., Harris, N., Lewis, S., Misra, S. and Shu, S. Q. (2005). FlyBase: genes and gene models. *Nucleic Acids Res,* 33 Database Issue, D390-395.

Etzold, T., Ulyanov, A. and Argos, P. (1996). SRS: information retrieval system for molecular biology data banks. *Methods Enzymol,* 266, 114-128.

Foster, I. (2003). The grid: computing without bounds. *Sci Am,* 288, 78-85.

Gattiker, A., Michoud, K., Rivoire, C., Auchincloss, A. H., Coudert, E., Lima, T., Kersey, P., Pagni, M., Sigrist, C. J., Lachaize, C., Veuthey, A. L., Gasteiger, E. and Bairoch, A. (2003). Automated annotation of microbial proteomes in SWISS-PROT. *Comput Biol Chem,* 27, 49-58.

Gough, J., Karplus, K., Hughey, R. and Chothia, C. (2001). Assignment of homology to genome sequences using a library of hidden Markov models that represent all proteins of known structure. *J Mol Biol,* 313, 903-919.

Haft, D. H., Selengut, J. D. and White, O. (2003). The TIGRFAMs database of protein families. *Nucleic Acids Res,* 31, 371-373.

Harris, M. A., Clark, J., Ireland, A., Lomax, J., Ashburner, M., Foulger, R., Eilbeck, K., Lewis, S., Marshall, B., Mungall, C., Richter, J., Rubin, G. M., Blake, J. A., Bult, C., Dolan, M., Drabkin, H., Eppig, J. T., Hill, D. P., Ni, L., Ringwald, M., Balakrishnan, R., Cherry, J. M., Christie, K. R., Costanzo, M. C., Dwight, S. S., Engel, S., Fisk, D. G., Hirschman, J. E., Hong, E. L., Nash, R. S., Sethuraman, A., Theesfeld, C. L., Botstein, D., Dolinski, K., Feierbach, B., Berardini, T., Mundodi, S., Rhee, S. Y., Apweiler, R., Barrell, D., Camon, E., Dimmer, E., Lee, V., Chisholm, R., Gaudet, P., Kibbe, W., Kishore, R., Schwarz, E. M., Sternberg, P., Gwinn, M., Hannick, L., Wortman, J., Berriman, M., Wood, V., de la Cruz, N., Tonellato, P., Jaiswal, P., Seigfried, T. and

White, R. (2004). The Gene Ontology (GO) database and informatics resource. *Nucleic Acids Res,* 32 Database issue, D258-261.

Hulo, N., Sigrist, C. J., Le Saux, V., Langendijk-Genevaux, P. S., Bordoli, L., Gattiker, A., De Castro, E., Bucher, P. and Bairoch, A. (2004). Recent improvements to the PROSITE database. *Nucleic Acids Res,* 32 Database issue, D134-137.

Hubbard, T., Andrews, D., Caccamo, M., Cameron, G., Chen, Y., Clamp, M., Clarke, L., Coates, G., Cox, T., Cunningham, F., Curwen, V., Cutts, T., Down, T., Durbin, R., Fernandez-Suarez, X. M., Gilbert, J., Hammond, M., Herrero, J., Hotz, H., Howe, K., Iyer, V., Jekosch, K., Kahari, A., Kasprzyk, A., Keefe, D., Keenan, S., Kokocinsci, F., London, D., Longden, I., McVicker, G., Melsopp, C., Meidl, P., Potter, S., Proctor, G., Rae, M., Rios, D., Schuster, M., Searle, S., Severin, J., Slater, G., Smedley, D., Smith, J., Spooner, W., Stabenau, A., Stalker, J., Storey, R., Trevanion, S., Ureta-Vidal, A., Vogel, J., White, S., Woodwark, C. and Birney, E. (2005). Ensembl 2005. *Nucleic Acids Res,* 33 Database Issue, D447-453.

Kanz, C., Aldebert, P., Althorpe, N., Baker, W., Baldwin, A., Bates, K., Browne, P., van den Broek, A., Castro, M., Cochrane, G., Duggan, K., Eberhardt, R., Faruque, N., Gamble, J., Diez, F. G., Harte, N., Kulikova, T., Lin, Q., Lombard, V., Lopez, R., Mancuso, R., McHale, M., Nardone, F., Silventoinen, V., Sobhany, S., Stoehr, P., Tuli, M. A., Tzouvara, K., Vaughan, R., Wu, D., Zhu, W. and Apweiler, R. (2005). The EMBL Nucleotide Sequence Database. *Nucleic Acids Res,* 33 Database Issue, D29-33.

Kasprzyk, A., Keefe, D., Smedley, D., London, D., Spooner, W., Melsopp, C., Hammond, M., Rocca-Serra, P., Cox, T. and Birney, E. (2004). EnsMart: a generic system for fast and flexible access to biological data. *Genome Res,* 14, 160-169.

Kersey, P., Bower, L., Morris, L., Horne, A., Petryszak, R., Kanz, C., Kanapin, A., Das, U., Michoud, K., Phan, I., Gattiker, A., Kulikova, T., Faruque, N., Duggan, K., McLaren, P., Reimholz, B., Duret, L., Penel, S., Reuter, I. and Apweiler, R. (2005). Integr8 and Genome Reviews: integrated views of complete genomes and proteomes. *Nucleic Acids Res,* 33 Database Issue, D297-302.

Kersey, P. J., Duarte, J., Williams, A., Karavidopoulou, Y., Birney, E. and Apweiler, R. (2004). The International Protein Index: an integrated database for proteomics experiments. *Proteomics,* 4, 1985-1988.

Kersey, P. J., Morris, L., Hermjakob, H. and Apweiler, R. (2003). Integr8: enhanced inter-operability of European molecular biology databases. *Methods Inf Med,* 42, 154-160.

Kriventseva, E. V., Servant, F. and Apweiler, R. (2003). Improvements to CluSTr: the database of SWISS-PROT+TrEMBL protein clusters. *Nucleic Acids Res,* 31, 388-389.

Leinonen, R., Diez, F. G., Binns, D., Fleischmann, W., Lopez, R. and Apweiler, R. (2004). UniProt archive. *Bioinformatics,* 20, 3236-3237.

Letunic, I., Copley, R. R., Schmidt, S., Ciccarelli, F. D., Doerks, T., Schultz, J., Ponting, C. P. and Bork, P. (2004). SMART 4.0: towards genomic data integration. *Nucleic Acids Res,* 32 Database issue, D142-144.

Lo Conte, L., Brenner, S. E., Hubbard, T. J., Chothia, C. and Murzin, A. G. (2002). SCOP database in 2002: refinements accommodate structural genomics. *Nucleic Acids Res,* 30, 264-267.

Mulder, N. J., Apweiler, R., Attwood, T. K., Bairoch, A., Bateman, A., Binns, D., Bradley, P., Bork, P., Bucher, P., Cerutti, L., Copley, R., Courcelle, E., Das, U., Durbin, R., Fleischmann, W., Gough, J., Haft, D., Harte, N., Hulo, N., Kahn, D., Kanapin, A., Krestyaninova, M., Lonsdale, D., Lopez, R., Letunic, I., Madera, M., Maslen, J., McDowall, J., Mitchell, A., Nikolskaya, A. N., Orchard, S., Pagni, M., Ponting, C. P., Quevillon, E., Selengut, J., Sigrist, C. J., Silventoinen, V., Studholme, D. J., Vaughan, R. and Wu, C. H. (2005). InterPro, progress and status in 2005. *Nucleic Acids Res,* 33 Database Issue, D201-205.

Pruess, M., Fleischmann, W., Kanapin, A., Karavidopoulou, Y., Kersey, P., Kriventseva, E., Mittard, V., Mulder, N., Phan, I., Servant, F. and Apweiler, R. (2003). The Proteome Analysis database: a tool for the in silico analysis of whole proteomes. *Nucleic Acids Res,* 31, 414-417.

Pruitt, K. D., Tatusova, T. and Maglott, D. R. (2005). NCBI Reference Sequence (RefSeq): a curated non-redundant sequence database of genomes, transcripts and proteins. *Nucleic Acids Res,* 33 Database Issue, D501-504.

Servant, F., Bru, C., Carrere, S., Courcelle, E., Gouzy, J., Peyruc, D. and Kahn, D. (2002). ProDom: automated clustering of homologous domains. *Brief Bioinform,* 3, 246-251.

Sherry, S. T., Ward, M. H., Kholodov, M., Baker, J., Phan, L., Smigielski, E. M. and Sirotkin, K. (2001). dbSNP: the NCBI database of genetic variation. *Nucleic Acids Res,* 29, 308-311.

Smith, T. F., Waterman, M. S. and Fitch, W. M. (1981). Comparative biosequence metrics. *J Mol Evol,* 18, 38-46.

Tateno, Y., Saitou, N., Okubo, K., Sugawara, H. and Gojobori, T. (2005). DDBJ in collaboration with mass-sequencing teams on annotation. *Nucleic Acids Res,* 33 Database Issue, D25-28.

Velankar, S., McNeil, P., Mittard-Runte, V., Suarez, A., Barrell, D., Apweiler, R. and Henrick, K. (2005). E-MSD: an integrated data resource for bioinformatics. *Nucleic Acids Res,* 33 Database Issue, D262-265.

Wu, C. H., Yeh, L. S., Huang, H., Arminski, L., Castro-Alvear, J., Chen, Y., Hu, Z., Kourtesis, P., Ledley, R. S., Suzek, B. E., Vinayaka, C. R., Zhang, J. and Barker, W. C. (2003). The Protein Information Resource. *Nucleic Acids Res,* 31, 345-347.

Index

C

D

J

K

L

M

T

U